TIME SERIES ANALYSIS AND APPLICATION

TIME SERIES ANALYSIS AND APPLICATION

Dr. M.N. Haque

RANDOM PUBLICATIONS
NEW DELHI - 110 002 (INDIA)

Time Series Analysis and Application

ISBN 978-93-51117-13-1

Published in 2015 in India by

RANDOM PUBLICATIONS

4376-A/4B, Gali Murari Lal, Ansari Road
New Delhi-110 002
Phone: +9111-43580356, 23289044
E-mail: randomexports@gmail.com; sales@randompublications.com; info@randompublications.com

Reprinted 2026

Type Setting by: Friends Media, Delhi-110089

Preface

A time series is a sequence of data points, measured typically at successive points in time spaced at uniform time intervals. Examples of time series are the daily closing value of the Dow Jones Industrial Average and the annual flow volume of the Nile River at Grand Ethiopian Renaissance Dam. Time series are very frequently plotted via line charts. Time series are used in statistics, signal processing, pattern recognition, econometrics, mathematical finance, weather forecasting, earthquake prediction, electroencephalography, control engineering, astronomy, communications engineering, and largely in any domain of applied science and engineering which involves temporal measurements. Time series analysis comprises methods for analysing time series data in order to extract meaningful statistics and other characteristics of the data. Time series forecasting is the use of a model to predict future values based on previously observed values. While regression analysis is often employed in such a way as to test theories that the current values of one or more independent time series affect the current value of another time series, this type of analysis of time series is not called "time series analysis", which focuses on comparing values of a single time series or multiple dependent time series at different points in time.

Time series data have a natural temporal ordering. This makes time series analysis distinct from cross-sectional studies, in which there is no natural ordering of the observations. Time series analysis is also distinct from spatial data analysis where the observations typically relate to geographical locations. A stochastic model for a time series will generally reflect the fact that observations close together in time will be more closely related than observations further apart. In addition, time series models will often make use of the natural one-way ordering of time so that values for a given period will be expressed as deriving in some way from past values, rather than from future values. Time series analysis can be applied to real-valued, continuous data, discrete numeric data, or discrete symbolic data. Methods for time series analyses may be divided into two classes: frequency-domain methods and time-domain methods. The former include spectral analysis and recently wavelet analysis; the latter include auto-correlation and cross-correlation analysis. In time domain correlation analyses can be made in a filter-like manner using scaled correlation, thereby mitigating the need to operate in frequency domain. Additionally, time series analysis techniques may be divided into parametric and non-parametric methods. The parametric approaches assume that the underlying stationary stochastic process has a certain structure which can be described using a small number of parameters. In these approaches, the task is to estimate the parameters of the model that describes the stochastic process. By contrast, non-parametric approaches explicitly estimate the covariance or the spectrum of the process without assuming that the process has any particular structure. Methods of time series analysis may also be divided into linear and non-linear, and univariate and multivariate. There are several types of motivation and data analysis available for time series which are appropriate for different purposes.

This book presents an accessible approach to understanding time series models and their applications.

I thank all members of my team who have helped in the preparation of the book. My special thanks go to "Random Publications" who have published the book.

— Dr. M.N. Haque

Contents

Chepter 1

Characteristics of Time Series

A time series is a sequence of data points, measured typically at successive points in time spaced at uniform time intervals. Examples of time series are the daily closing value of the Dow Jones Industrial Average and the annual flow volume of the Nile River at Grand Ethiopian Renaissance Dam. Time series are very frequently plotted via line charts. Time series are used in statistics, signal processing, pattern recognition, econometrics, mathematical finance, weather forecasting, earthquake prediction, electroencephalography, control engineering, astronomy, communications engineering, and largely in any domain of applied science and engineering which involves temporal measurements.

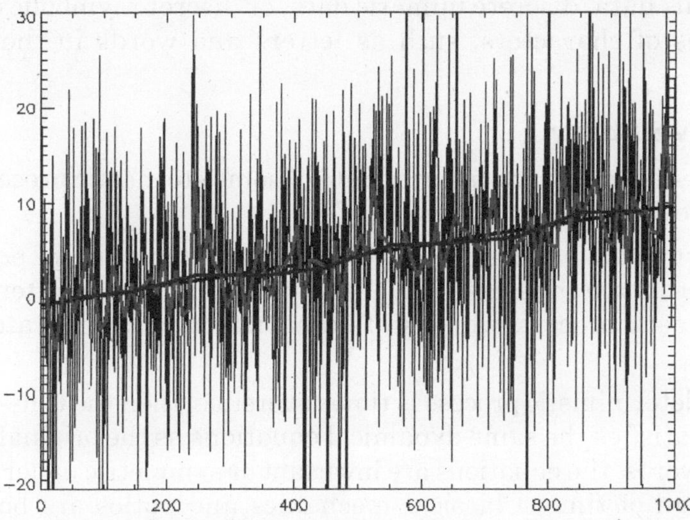

***Figure:** Time series: random data plus trend, with best-fit line and different applied filters*

Time series analysis comprises methods for analyzing time series data in order to extract meaningful statistics and other characteristics of the data. Time series forecasting is the use of a model to predict future values based on previously observed values. While regression analysis is often employed in such a way as to test theories that the current values of one or more independent time series affect the current value of another time series, this type of analysis of time series is not called "time series analysis", which focuses on comparing values of a single time series or multiple dependent time series at different points in time.

Time series data have a natural temporal ordering. This makes time series analysis distinct from cross-sectional studies, in which there is no natural ordering of the observations (e.g. explaining people's wages by reference to their respective education levels, where the individuals' data could be entered in any order). Time series analysis is also distinct from spatial data analysis where the observations typically relate to geographical locations (e.g. accounting for house prices by the location as well as the intrinsic characteristics of the houses). A stochastic model for a time series will generally reflect the fact that observations close together in time will be more closely related than observations further apart. In addition, time series models will often make use of the natural one-way ordering of time so that values for a given period will be expressed as deriving in some way from past values, rather than from future values Time series analysis can be applied to real-valued, continuous data, discrete numeric data, or discrete symbolic data (i.e. sequences of characters, such as letters and words in the English language.

Time Reversibility

Time reversibility is an attribute of some stochastic processes and some deterministic processes.

If a stochastic process is time reversible, then it is not possible to determine, given the states at a number of points in time after running the stochastic process, which state came first and which state arrived later.

If a deterministic process is time reversible, then the time-reversed process satisfies the same dynamical equations as the original process; in other words, the equations are invariant or symmetric under a change in the sign of time. Classical mechanics and optics are both time-reversible. Modern physics is not quite time-reversible; instead it exhibits a broader symmetry, CPT symmetry.

Time reversibility generally occurs when, within a process, it can be broken up into sub-processes which undo the effects of each other. For example, in phylogenetics, a time-reversible nucleotide substitution model such as the generalised time reversible model has the total overall rate into a certain nucleotide equal to the total rate out of that same nucleotide.

Time Reversal, specifically in the field of acoustics, is a process by which the linearity of sound waves is used to reverse a received signal; this signal is then re-emitted and a temporal compression occurs, resulting in a reverse of the initial excitation waveform being played at the initial source. Mathias Fink is credited with proving Acoustic Time Reversal by experiment.

Stochastic Processes

A formal definition of time-reversibility is stated by Tong in the context of time-series. In general, a Gaussian process is time-reversible. The process defined by a time-series model which represents values as a linear combination of past values and of present and past innovations is, except for limited special cases, not time-reversible unless the innovations have a normal distribution (in which case the model is a Gaussian process).

A stationary Markov Chain is reversible if the transition matrix $\{p_{ij}\}$ and the stationary distribution $\{\pi_j\}$ satisfy

$$\pi_i p_{ij} = \pi_j p_{ji},$$

for all i and j. Such Markov Chains provide examples of stochastic processes which are time-reversible but non-Gaussian.

Time reversal of numerous classes of stochastic processes have been studied including Lévy processes stochastic networks (Kelly's lemma) birth and death processes Markov chains and piecewise deterministic Markov processes.

Methods for Time Series Analyses

Methods for time series analyses may be divided into two classes: frequency-domain methods and time-domain methods. The former include spectral analysis and recently wavelet analysis; the latter include auto-correlation and cross-correlation analysis. In time domain correlation analyses can be made in a filter-like manner using scaled correlation, thereby mitigating the need to operate in frequency domain.

Additionally, time series analysis techniques may be divided into parametric and non-parametric methods. The parametric approaches assume that the underlying stationary stochastic process has a certain structure which can be described using a small number of parameters (for example, using an autoregressive or moving average model). In these approaches, the task is to estimate the parameters of the model that describes the stochastic process. By contrast, non-parametric approaches explicitly estimate the covariance or the spectrum of the process without assuming that the process has any particular structure.

Methods of time series analysis may also be divided into linear and non-linear, and univariate and multivariate.

Analysis

There are several types of motivation and data analysis available for time series which are appropriate for different purposes.

Motivation

The context of statistics, econometrics, quantitative finance, seismology, meteorology, and geophysics the primary goal of time series analysis is forecasting. In the context of signal processing, control engineering and communication engineering it is used for signal detection and estimation, while in the context of data mining, pattern recognition and machine learning time series analysis can be used for clustering, classification, query by content, anomaly detection as well as forecasting.

Exploratory Analysis

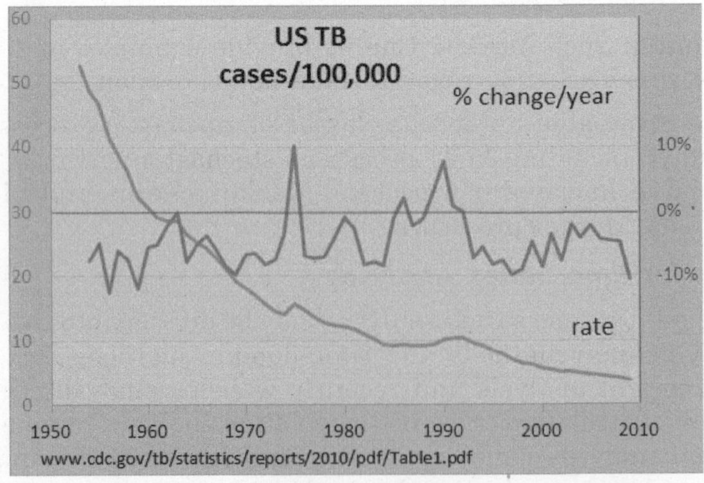

Figure: *Tuberculosis incidence US 1953-2009*

The clearest way to examine a regular time series manually is with a line chart such as the one shown for tuberculosis in the United

States, made with a spreadsheet program. The number of cases was standardized to a rate per 100,000 and the percent change per year in this rate was calculated. The nearly steadily dropping line shows that the TB incidence was decreasing in most years, but the percent change in this rate varied by as much as +/- 10%, with 'surges' in 1975 and around the early 1990s. The use of both vertical axes allows the comparison of two time series in one graphic. Other techniques include:

1. Autocorrelation analysis to examine serial dependence
2. Spectral analysis to examine cyclic behaviour which need not be related to seasonality. For example, sun spot activity varies over 11 year cycles. Other common examples include celestial phenomena, weather patterns, neural activity, commodity prices, and economic activity.
3. Separation into components representing trend, seasonality, slow and fast variation, and cyclical irregularity:

Prediction and Forecasting

- Fully formed statistical models for stochastic simulation purposes, so as to generate alternative versions of the time series, representing what might happen over non-specific time-periods in the future
- Simple or fully formed statistical models to describe the likely outcome of the time series in the immediate future, given knowledge of the most recent outcomes (forecasting).
- Forecasting on time series is usually done using automated statistical software packages and programming languages, such as R, S, SAS, SPSS, Minitab, Pandas (Python) and many others.

Statistical Classification

In machine learning and statistics, classification is the problem of identifying to which of a set of categories (sub-populations) a new observation belongs, on the basis of a training set of data containing observations (or instances) whose category membership is known. An example would be assigning a given email into "spam" or "non-spam" classes or assigning a diagnosis to a given patient as described by observed characteristics of the patient (gender, blood pressure, presence or absence of certain symptoms, etc.).

In the terminology of machine learning, classification is considered an instance of supervised learning, i.e. learning where a training set of correctly identified observations is available. The corresponding unsupervised procedure is known as clustering, and involves grouping data into categories based on some measure of inherent similarity or distance. Often, the individual observations are analyzed into a set of quantifiable properties, known variously explanatory variables, features, etc. These properties may variously be categorical (e.g. "A", "B", "AB" or "O", for blood type), ordinal (e.g. "large", "medium" or "small"), integer-valued (e.g. the number of occurrences of a part word in an email) or real-valued (e.g. a measurement of blood pressure). Other classifiers work by comparing observations to previous observations by means of a similarity or distance function.

An algorithm that implements classification, especially in a concrete implementation, is known as a classifier. The term "classifier" sometimes also refers to the mathematical function, implemented by a classification algorithm, that maps input data to a category.

Terminology across fields is quite varied. In statistics, where classification is often done with logistic regression or a similar procedure, the properties of observations are termed explanatory variables (or independent variables, regressors, etc.), and the categories to be predicted are known as outcomes, which are considered to be possible values of the dependent variable. In machine learning, the observations are often known as instances, the explanatory variables are termed features (grouped into a feature vector), and the possible categories to be predicted are classes. There is also some argument over whether classification methods that do not involve a statistical model can be considered "statistical".

Relation to Other Problems

Classification and clustering are examples of the more general problem of pattern recognition, which is the assignment of some sort of output value to a given input value. Other examples are regression, which assigns a real-valued output to each input; sequence labelling, which assigns a class to each member of a sequence of values (for example, part of speech tagging, which assigns a part of speech to each word in an input sentence); parsing, which assigns a parse tree to an input sentence, describing the syntactic structure of the sentence; etc.

A common subclass of classification is probabilistic classification. Algorithms of this nature use statistical inference to find the best class for a given instance. Unlike other algorithms, which simply output a "best" class, probabilistic algorithms output a probability of the instance being a member of each of the possible classes. The best class is normally then selected as the one with the highest probability. However, such an algorithm has numerous advantages over non-probabilistic classifiers:

- It can output a confidence value associated with its choice (in general, a classifier that can do this is known as a confidence-weighted classifier).
- Correspondingly, it can abstain when its confidence of choosing any particular output is too low.
- Because of the probabilities which are generated, probabilistic classifiers can be more effectively incorporated into larger machine-learning tasks, in a way that partially or completely avoids the problem of error propagation.

Frequentist Procedures

Early work on statistical classification was undertaken by Fisher, in the context of two-group problems, leading to Fisher's linear discriminant function as the rule for assigning a group to a new observation. This early work assumed that data-values within each of the two groups had a multivariate normal distribution. The extension of this same context to more than two-groups has also been considered with a restriction imposed that the classification rule should be linear. Later work for the multivariate normal distribution allowed the classifier to be nonlinear: several classification rules can be derived based on slight different adjustments of the Mahalanobis distance, with a new observation being assigned to the group whose centre has the lowest adjusted distance from the observation.

Bayesian Procedures

Unlike frequentist procedures, Bayesian classification procedures provide a natural way of taking into account any available information about the relative sizes of the sub-populations associated with the different groups within the overall population. Bayesian procedures tend to be computationally expensive and, in the days before Markov chain Monte Carlo computations were developed, approximations for Bayesian clustering rules were devised.

Some Bayesian procedures involve the calculation of group membership probabilities: these can be viewed as providing a more informative outcome of a data analysis than a simple attribution of a single group-label to each new observation.

Binary and Multiclass Classification

Classification can be thought of as two separate problems – binary classification and multiclass classification. In binary classification, a better understood task, only two classes are involved, whereas multiclass classification involves assigning an object to one of several classes. Since many classification methods have been developed specifically for binary classification, multiclass classification often requires the combined use of multiple binary classifiers.

Feature Vectors

Most algorithms describe an individual instance whose category is to be predicted using a feature vector of individual, measurable properties of the instance. Each property is termed a feature, also known in statistics as an explanatory variable (or independent variable, although in general different features may or may not be statistically independent). Features may variously be binary ("male" or "female"); categorical (e.g. "A", "B", "AB" or "O", for blood type); ordinal (e.g. "large", "medium" or "small"); integer-valued (e.g. the number of occurrences of a particular word in an email); or real-valued (e.g. a measurement of blood pressure). If the instance is an image, the feature values might correspond to the pixels of an image; if the instance is a piece of text, the feature values might be occurrence frequencies of different words. Some algorithms work only in terms of discrete data and require that real-valued or integer-valued data be discretized into groups (e.g. less than 5, between 5 and 10, or greater than 10).

The vector space associated with these vectors is often called the feature space. In order to reduce the dimensionality of the feature space, a number of dimensionality reduction techniques can be employed.

Linear Classifiers

A large number of algorithms for classification can be phrased in terms of a linear function that assigns a score to each possible category k by combining the feature vector of an instance with a vector of weights, using a dot product. The predicted category is the one with the highest score. This type of score function is known as a linear predictor function and has the following general form:

$$\text{score}(\mathbf{X}_i, k) = \boldsymbol{\beta}_k \cdot \mathbf{X}_i,$$

where X_i is the feature vector for instance i, β_k is the vector of weights corresponding to category k, and score(X_i, k) is the score associated with assigning instance i to category k. In discrete choice theory, where instances represent people and categories represent choices, the score is considered the utility associated with person i choosing category k.

Algorithms with this basic setup are known as linear classifiers. What distinguishes them is the procedure for determining (training) the optimal weights/coefficients and the way that the score is interpreted.

Probit Model

In statistics, a probit model is a type of regression where the dependent variable can only take two values, for example married or not married. The name is from probability + unit. The purpose of the model is to estimate the probability that an observation with particular characteristics will fall into a specific one of the categories; moreover, if estimated probabilities greater than 1/2 are treated as classifying an observation into a predicted category, the probit model is a type of binary classification model. A probit model is a popular specification for an ordinal or a binary response model. As such it treats the same set of problems as does logistic regression using similar techniques. The probit model, which employs a probit link function, is most often estimated using the standard maximum likelihood procedure, such an estimation being called a probit regression.

Conceptual Framework

Suppose response variable Y is binary, that is it can have only two possible outcomes which we will denote as 1 and 0. For example Y may represent presence/absence of a certain condition, success/failure of some device, answer yes/no on a survey, etc. We also have a vector of regressors X, which are assumed to influence the outcome Y. Specifically, we assume that the model takes the form

$$\Pr(Y=1 \mid X) = \Phi(X'\beta),$$

where Pr denotes probability, and ϑ is the Cumulative Distribution Function (CDF) of the standard normal distribution. The parameters β are typically estimated by maximum likelihood. It is possible to motivate the probit model as a latent variable model. Suppose there exists an auxiliary random variable

$$Y^* = X'\beta + \varepsilon,$$

where $\varepsilon \sim N(0, 1)$. Then Y can be viewed as an indicator for whether this latent variable is positive:

$$Y = \begin{cases} 1 & \text{if } Y^* > 0 \text{ i.e. } -\varepsilon < X'\beta, \\ 0 & \text{otherwise.} \end{cases}$$

The use of the standard normal distribution causes no loss of generality compared with using an arbitrary mean and standard deviation because adding a fixed amount to the mean can be compensated by subtracting the same amount from the intercept, and multiplying the standard deviation by a fixed amount can be compensated by multiplying the weights by the same amount.

To see that the two models are equivalent, note that

$$\begin{aligned}\Pr(Y=1\mid X) &= \Pr(Y^*>0)=\Pr(X'\beta+\varepsilon>0)\\ &= \Pr(\varepsilon>-X'\beta)\\ &= \Pr(\varepsilon<X'\beta) \quad \text{(by symmetry of the normal dist)}\\ &= \Phi(X'\beta)\end{aligned}$$

Model Estimation

Maximum Likelihood Estimation

Suppose data set $\{y_i, x_i\}_{i=1}^n$ contains n independent statistical units corresponding to the model above. Then their joint log-likelihood function is

$$\ln \mathcal{L}(\beta)=\sum_{i=1}^{n}\Big(y_i \ln \Phi(x_{i'}\beta)+(1-y_i)\ln\big(1-\Phi(x_{i'}\beta)\big)\Big)$$

The estimator $\hat{\beta}$ which maximizes this function will be consistent, asymptotically normal and efficient provided that E[XX'] exists and is not singular. It can be shown that this log-likelihood function is globally concave in β, and therefore standard numerical algorithms for optimization will converge rapidly to the unique maximum.

Asymptotic distribution for $\hat{\beta}$ is given by

$$\sqrt{n}(\hat{\beta}-\beta)\xrightarrow{d}\mathcal{N}(0,\Omega^{-1}),$$

where

$$\Omega=\mathrm{E}\Big[\frac{\varphi^2(X'\beta)}{\Phi(X'\beta)(1-\Phi(X'\beta))}XX'\Big], \qquad \hat{\Omega}=\frac{1}{n}\sum_{i=1}^{n}\frac{\varphi^2(x_i'\hat{\beta})}{\Phi(x_i'\hat{\beta})(1-\Phi(x_i'\hat{\beta}))}x_i x_i'$$

and φ = ϑ' is the Probability Density Function (PDF) of standard normal distribution.

Berkson's Minimum Chi-Square Method

This method can be applied only when there are many observations of response variable having the same value of the vector of regressors x_i (such situation may be referred to as "many observations per cell"). More specifically, the model can be formulated as follows.

Suppose among n observations $\{y_i, x_i\}_{i=1}^n$ there are only T distinct values of the regressors, which can be denoted as $\{x_{(1)},\ldots,x_{(T)}\}$. Let be n_t the number of observations with $x_i = x_{(t)}$, and r_t the number of such observations with $y_i = 1$. We assume that there are indeed "many" observations per each "cell": for each $t, \lim_{n\to\infty} n_t / n = c_t > 0$.

Denote

$$\hat{p}_t = r_t / n_t$$

$$\hat{\sigma}_t^2 = \frac{1}{n_t} \frac{\hat{p}_t(1-\hat{p}_t)}{\varphi^2\left(\Phi^{-1}(\hat{p}_t)\right)}$$

Then Berkson's minimum chi-square estimator is a generalized least squares estimator in a regression of $\Phi^{-1}(\hat{p}_t)$ on $x_{(t)}$ with weights:

$$\hat{\beta} = \Big(\sum_{t=1}^{T} \hat{\sigma}_t^{-2} x_{(t)} x'_{(t)}\Big)^{-1} \sum_{t=1}^{T} \hat{\sigma}_t^{-2} x_{(t)} \Phi^{-1}(\hat{p}_t)$$

It can be shown that this estimator is consistent (as n→∞ and T fixed), asymptotically normal and efficient. Its advantage is the presence of a closed-form formula for the estimator. However, it is only meaningful to carry out this analysis when individual observations are not available, only their aggregated counts r_t, n_t , $x_{(t)}$ and (for example in the analysis of voting behaviour).

Gibbs Sampling

Gibbs sampling of a probit model is possible because regression models typically use normal prior distributions over the weights, and this distribution is conjugate with the normal distribution of the errors (and hence of the latent variablesY*). The model can be described as

$$\begin{aligned}
&\boldsymbol{\beta} && \sim \mathcal{N}(\mathbf{b}_0, \mathbf{B}_0) \\
&y_i^* \mid \mathbf{x}_i, \boldsymbol{\beta} && \sim \mathcal{N}(\mathbf{x}_i'\boldsymbol{\beta}, 1) \\
&y_i && = \begin{cases} 1 & \text{if } y_i^* > 0 \\ 0 & \text{otherwise} \end{cases}
\end{aligned}$$

From this, we can determine the full conditional densities needed:

$$\begin{aligned}
&\mathbf{B} && = (\mathbf{B}_0^{-1} + \mathbf{X}'\mathbf{X})^{-1} \\
&\boldsymbol{\beta} \mid \mathbf{y}^* && \sim \mathcal{N}(\mathbf{B}(\mathbf{B}_0^{-1}\mathbf{b}_0 + \mathbf{X}'\mathbf{y}^*), \mathbf{B}) \\
&y_i^* \mid y_i = 0, \mathbf{x}_i, \boldsymbol{\beta} && \sim \mathcal{N}(\mathbf{x}_i'\boldsymbol{\beta}, 1)[y_i^* < 0] \\
&y_i^* \mid y_i = 1, \mathbf{x}_i, \boldsymbol{\beta} && \sim \mathcal{N}(\mathbf{x}_i'\boldsymbol{\beta}, 1)[y_i^* \geq 0]
\end{aligned}$$

The result for β is given in the article on Bayesian linear regression, although specified with different notation.

The only trickiness is in the last two equations. The notation $[y_i^* < 0]$ is the Iverson bracket, sometimes written or $\mathcal{I}(y_i^* < 0)$ similar. It indicates that the distribution must be truncated within the given range, and rescaled appropriately. In this particular case, a truncated normal distribution arises. Sampling from this distribution depends on how much is truncated. If a large fraction of the original mass remains, sampling can be easily done with rejection sampling — simply sample a number from the non-truncated distribution, and reject it if it falls outside the restriction imposed by the truncation. If sampling from only a small fraction of the original mass, however (e.g. if sampling from one of the tails of the normal distribution — for example if is around 3 or more, and a negative sample is desired), then this will be inefficient and it becomes necessary to fall back on other sampling algorithms. General sampling from the truncated normal can be achieved using approximations to the normal CDF and the probit function, and R has a function rtnorm() for generating truncated-normal samples.

Model Evaluation

The suitability of an estimated binary model can be evaluated by counting the number of true observations equaling 1, and the number equaling zero, for which the model assigns a correct predicted classification by treating any estimated probability above 1/2 (or, below 1/2), as an assignment of a prediction of 1 (or, of 0).

The Perceptron Algorithm

In the modern sense, the perceptron is an algorithm for learning a binary classifier: a function that maps its input *x* (a real-valued vector) to an output value *f(x)* (a single binary value):

$$f(x) = \begin{cases} 1 & \text{if } w \cdot x + b > 0 \\ 0 & \text{otherwise} \end{cases}$$

where *w* is a vector of real-valued weights, $w \cdot x$ is the dot product (which here computes a weighted sum), and *b* is the 'bias', a constant term that does not depend on any input value.

The value of $f(x)$ (0 or 1) is used to classify as either a positive or a negative instance, in the case of a binary classification problem. If is negative, then the weighted combination of inputs must produce a positive value greater than $|b|$ in order to push the classifier neuron over the 0 threshold. Spatially, the bias alters the position (though not the orientation) of the decision boundary. The perceptron learning algorithm does not terminate if the learning set is not linearly separable. If the vectors are not linearly separable learning will never reach a point where all vectors are classified properly. The most famous example of the perceptron's inability to solve problems with linearly nonseparable vectors is the Boolean exclusive-or problem. The solution spaces of decision boundaries for all binary functions and learning behaviours are studied in the reference.

In the context of neural networks, a perceptron is an artificial neuron using the Heaviside step function as the activation function. The perceptron algorithm is also termed the single-layer perceptron, to distinguish it from a multilayer perceptron, which is a misnomer for a more complicated neural network. As a linear classifier, the single-layer perceptron is the simplest feedforward neural network.

Support Vector Machines

More formally, a support vector machine constructs a hyperplane or set of hyperplanes in a high- or infinite-dimensional space, which can be used for classification, regression, or other tasks. Intuitively, a good separation is achieved by the hyperplane that has the largest distance to the nearest training data point of any class (so-called functional margin), since in general the larger the margin the lower the generalization error of the classifier.

Whereas the original problem may be stated in a finite dimensional space, it often happens that the sets to discriminate are not linearly separable in that space. For this reason, it was proposed that the original finite-dimensional space be mapped into a much higher-dimensional space, presumably making the separation easier in that space. To keep the computational load reasonable, the mappings used by SVM schemes are designed to ensure that dot products may be computed easily in terms of the variables in the original space, by defining them in terms of a kernel function $k(x, y)$ selected to suit the problem.

The hyperplanes in the higher-dimensional space are defined as the set of points whose dot product with a vector in that space is constant. The vectors defining the hyperplanes can be chosen to be linear combinations with parameters α_i of images of feature vectors that occur in the data base. With this choice of a hyperplane, the points x in the feature space that are mapped into the hyperplane are defined by the relation: $\sum_i \alpha_i k(x_i, x) = \text{constant}$. Note that if $k(x, y)$ becomes small as grows further away from x, each term in the sum measures the degree of closeness of the test point to the corresponding data base point x_i. In this way, the sum of kernels above can be used to measure the relative nearness of each test point to the data points originating in one or the other of the sets to be discriminated. Note the fact that the set of points x mapped into any hyperplane can be quite convoluted as a result, allowing much more complex discrimination between sets which are not convex at all in the original space.

Linear Discriminant Analysis.

Linear discriminant analysis $\vec{x}$ (LDA) and the related Fisher's linear discriminant are methods used in statistics, pattern recognition and machine learning to find a linear combination of features which characterizes or separates two or more classes of objects or events. The resulting combination may be used as a linear classifier, or, more commonly, for dimensionality reduction before later classification.

LDA is closely related to ANOVA (analysis of variance) and regression analysis, which also attempt to express one dependent variable as a linear combination of other features or measurements. However, ANOVA uses categorical independent variables and a continuous dependent variable, whereas discriminant analysis has continuous independent variables and a categorical dependent variable (i.e. the class label). Logistic regression and probit regression are more similar to LDA, as they also explain a categorical variable by the values of continuous independent variables. These other methods are preferable in applications where it is not reasonable to assume that the independent variables are normally distributed, which is a fundamental assumption of the LDA method.

LDA is also closely related to principal component analysis (PCA) and factor analysis in that they both look for linear combinations of variables which best explain the data. LDA explicitly attempts to model the difference between the classes of data. PCA on the other hand does not take into account any difference in class, and factor analysis builds the feature combinations based on differences rather than similarities. Discriminant analysis is also different from factor analysis in that it is not an interdependence technique: a distinction between independent variables and dependent variables (also called criterion variables) must be made.

LDA works when the measurements made on independent variables for each observation are continuous quantities. When dealing with categorical independent variables, the equivalent technique is discriminant correspondence analysis.

LDA for Two Classes

Consider a set of observations $\vec{x}$ (also called features, attributes, variables or measurements) for each sample of an object or event with known class y. This set of samples is called the training set. The classification problem is then to find a good predictor for the class y of any sample of the same distribution (not necessarily from the training set) given only an observation $\vec{x}$.

LDA approaches the problem by assuming that the conditional probability density functions $p(\vec{x} \mid y=0)$ and $p(\vec{x} \mid y=1)$ are both normally distributed with mean and covariance parameters $\left(\vec{\mu}_0, \Sigma_0\right)$ and $\left(\vec{\mu}_1, \Sigma_1\right)$, respectively. Under this assumption, the Bayes optimal solution is to predict points as being from the second class if the log of the likelihood ratios is below some threshold T, so that;

Without any further assumptions, the resulting classifier is referred to as QDA (quadratic discriminant analysis).

LDA instead makes the additional simplifying homoscedasticity assumption (i.e. that the class covariances are identical, so $\Sigma_0 = \Sigma_1 = \Sigma$) and that the covariances have full rank. In this case, several terms cancel and the above decision criterion becomes a threshold on the dot product

$$\vec{w} \cdot \vec{x} > c$$

for some threshold constant c, where

$$\vec{w} \propto \Sigma^{-1}(\vec{\mu}_1 - \vec{\mu}_0)$$

This means that the criterion of an input $\vec{x}$ being in a class y is purely a function of this linear combination of the known observations.

It is often useful to see this conclusion in geometrical terms: the criterion of an input $\vec{x}$ being in a class y is purely a function of projection of multidimensional-space point $\vec{x}$ onto vector $\vec{w}$ (thus, we only consider its direction). In other words, the observation belongs to y if corresponding is located on a certain side of a hyperplane perpendicular to $\vec{w}$. The location of the plane is defined by the threshold c.

Evaluation

Classifier performance depends greatly on the characteristics of the data to be classified. There is no single classifier that works best on all given problems (a phenomenon that may be explained by the no-free-lunch theorem). Various empirical tests have been performed to compare classifier performance and to find the characteristics of data that determine classifier performance. Determining a suitable classifier for a given problem is however still more an art than a science.

The measures precision and recall are popular metrics used to evaluate the quality of a classification system. More recently, receiver operating characteristic (ROC) curves have been used to evaluate the tradeoff between true- and false-positive rates of classification algorithms.

As a performance metric, the uncertainty coefficient has the advantage over simple accuracy in that it is not affected by the relative sizes of the different classes. Further, it will not penalize an algorithm for simply rearranging the classes.

Regression Analysis (Method of Prediction)

- Estimating future value of a signal based on its previous behaviour, e.g. predict the price of AAPL stock based on its previous price movements for that hour, day or month, or predict position of Apollo 11 spacecraft at a certain future moment based on its current trajectory (i.e. time series of its previous locations).
- Regression analysis is usually based on statistical interpretation of time series properties in time domain, pioneered by statisticians George Box and Gwilym Jenkins in the 50s.

Signal Estimation

- This approach is based on harmonic analysis and filtering of signals in the frequency domain using the Fourier transform, and spectral density estimation, the development of which was significantly accelerated during World War II by mathematician Norbert Wiener, electrical engineers Rudolf E. Kálmán, Dennis Gabor and others for filtering signals from noise and predicting signal values at a certain point in time.

Segmentation

- Splitting a time-series into a sequence of segments. It is often the case that a time-series can be represented as a sequence of individual segments, each with its own characteristic properties. For example, the audio signal from a conference call can be partitioned into pieces corresponding to the times during which each person was speaking. In time-series segmentation, the goal is to identify the segment boundary points in the time-series, and to characterize the dynamical properties associated with each segment. One can approach this problem using change-point detection, or by modelling the time-series as a more sophisticated system, such as a Markov jump linear system.

Models

Models for time series data can have many forms and represent different stochastic processes. When modelling variations in the level of a process, three broad classes of practical importance are the autoregressive (AR) models, the integrated (I) models, and the moving average (MA) models. These three classes depend linearly on previous data points. Combinations of these ideas produce autoregressive moving average (ARMA) and autoregressive integrated moving average (ARIMA) models. The autoregressive fractionally integrated moving average (ARFIMA) model generalizes the former three.

Extensions of these classes to deal with vector-valued data are available under the heading of multivariate time-series models and sometimes the preceding acronyms are extended by including an initial "V" for "vector", as in VAR for vector autoregression. An additional set of extensions of these models is available for use where the observed time-series is driven by some "forcing" time-series (which may not have a causal effect on the observed series): the distinction from the multivariate case is that the forcing series may be deterministic or under the experimenter's control. For these models, the acronyms are extended with a final "X" for "exogenous".

Non-linear dependence of the level of a series on previous data points is of interest, partly because of the possibility of producing a chaotic time series. However, more importantly, empirical investigations can indicate the advantage of using predictions derived from non-linear models, over those from linear models, as for example in nonlinear autoregressive exogenous models.

Among other types of non-linear time series models, there are models to represent the changes of variance over time (heteroskedasticity). These models represent autoregressive conditional heteroskedasticity (ARCH) and the collection comprises a wide variety of representation (GARCH, TARCH, EGARCH, FIGARCH, CGARCH, etc.). Here changes in variability are related to, or predicted by, recent past values of the observed series.

This is in contrast to other possible representations of locally varying variability, where the variability might be modelled as being driven by a separate time-varying process, as in a doubly stochastic model.

In recent work on model-free analyses, wavelet transform based methods (for example locally stationary wavelets and wavelet decomposed neural networks) have gained favour. Multiscale (often referred to as multiresolution) techniques decompose a given time series, attempting to illustrate time dependence at multiple scales.

A Hidden Markov model (HMM) is a statistical Markov model in which the system being modelled is assumed to be a Markov process with unobserved (hidden) states. An HMM can be considered as the simplest dynamic Bayesian network. HMM models are widely used in speech recognition, for translating a time series of spoken words into text.

Notation

A number of different notations are in use for time-series analysis. A common notation specifying a time series X that is indexed by the natural numbers is written

$$X = \{X_1, X_2, ...\}.$$

Another common notation is

$$Y = \{Y_t: t \in T\},$$

where T is the index set.

Conditions

There are two sets of conditions under which much of the theory is built:

Stationary Process

In mathematics and statistics, a stationary process (or strict(ly) stationary process or strong(ly) stationary process) is a stochastic process whose joint probability distribution does not change when shifted in time. Consequently, parameters such as the mean and variance, if they are present, also do not change over time and do not follow any trends.

Stationarity is used as a tool in time series analysis, where the raw data is often transformed to become stationary; for example, economic data are often seasonal and/or dependent on a non-stationary price level. An important type of non-stationary process that does not include a trend-like behaviour is the cyclostationary process.

Note that a "stationary process" is not the same thing as a "process with a stationary distribution". Indeed there are further possibilities for confusion with the use of "stationary" in the context of stochastic processes; for example a "time-homogeneous" Markov chain is sometimes said to have "stationary transition probabilities". Besides, all stationary Markov random processes are time-homogeneous.

Ergodic Process

In econometrics and signal processing, a stochastic process is said to be ergodic if its statistical properties (such as its mean and variance) can be deduced from a single, sufficiently long sample (realisation) of the process. The reasoning behind this is that any sample from the process must represent the average statistical properties of the entire process, so that regardless what sample you choose, it represents the whole process, and not just that section of the process. A process that changes erratically at an inconsistent rate is not said to be ergodic.

One can discuss the ergodicity of various properties of a stochastic process. For example, a wide-sense stationary process $X(t)$ has mean:

$$\mu_X(t) = E[X(t)] = \mu_X,$$

and autocovariance

$$r_X(\tau) = E[(X(t) - \mu_X(t))(X(t+\tau) - \mu_X(t+\tau))],$$

which do not change with time t.

One way to estimate the mean is to perform a time average:

$$\hat{\mu}_X(t)_T = \frac{1}{T}\int_0^T X(t)dt.$$

If $\hat{\mu}_X(t)_T$ converges in squared mean to the true value $\mu_X(t)$ as $T \to \infty$, then the process $X(t)$ is said to be mean-ergodic or mean-square ergodic in the first moment.

Likewise, one can estimate the autocovariance $r_x(\tau)$ by performing a time average:

Again, if $\hat{r}_X(\tau)_T$ converges in squared mean to the true autocovariance $r_X(\tau) = E[(X(t+\tau) - \mu_X(t+\tau))(X(t) - \mu_X(t))]$, then the process is said to be autocovariance-ergodic or mean-square ergodic in the second moment.

A process which is ergodic in the first and second moments is sometimes called ergodic in the wide sense.

An important example of an ergodic processes is the stationary Gaussian process with continuous spectrum.

However, ideas of stationarity must be expanded to consider two important ideas: strict stationarity and second-order stationarity. Both models and applications can be developed under each of these conditions, although the models in the latter case might be considered as only partly specified.

In addition, time-series analysis can be applied where the series are seasonally stationary or non-stationary. Situations where the amplitudes of frequency components change with time can be dealt with in time-frequency analysis which makes use of a time–frequency representation of a time-series or signal.

Models

The general representation of an autoregressive model, well known as AR(p), is

$$Y_t = \alpha_0 + \alpha_1 Y_{t-1} + \alpha_2 Y_{t-2} + \cdots + \alpha_p Y_{t-p} + \varepsilon_t$$

where the term ε_t is the source of randomness and is called white noise. It is assumed to have the following characteristics:

- $E[\varepsilon_t] = 0,$
- $E[\varepsilon_t^2] = \sigma^2,$
- $E[\varepsilon_t \varepsilon_s] = 0 \quad$ for all $t \neq s$.

With these assumptions, the process is specified up to second-order moments and, subject to conditions on the coefficients, may be second-order stationary.

If the noise also has a normal distribution, it is called normal or Gaussian white noise. In this case, the AR process may be strictly stationary, again subject to conditions on the coefficients.

Tools for investigating time-series data include:

- Consideration of the autocorrelation function and the spectral density function (also cross-correlation functions and cross-spectral density functions)
- Scaled cross- and auto-correlation functions to remove contributions of slow components
- Performing a Fourier transform to investigate the series in the frequency domain
- Use of a filter to remove unwanted noise
- Principal component analysis (or empirical orthogonal function analysis)
- Singular spectrum analysis
- "Structural" models:

- General State Space Models
- Unobserved Components Models

- Machine Learning
  - Artificial neural networks
  - Support Vector Machine
  - Fuzzy Logic
  - Gaussian Processes
- Hidden Markov model
- Control chart
  - Shewhart individuals control chart
  - CUSUM chart
  - EWMA chart
- Detrended fluctuation analysis
- Dynamic time warping
- Dynamic Bayesian network
- Time-frequency analysis techniques:
  - Fast Fourier Transform
  - Continuous wavelet transform
  - Short-time Fourier transform
  - Chirplet transform
  - Fractional Fourier transform
- Similarity measures:
  - Dynamic Time Warping
  - Hidden Markov Models
  - Edit distance
  - Total correlation
  - Newey–West estimator
  - Prais–Winsten transformation
  - Data as Vectors in a Metrizable Space
    - Minkowski distance
    - Mahalanobis distance
  - Data as Time Series with Envelopes
    - Global Standard Deviation
    - Local Standard Deviation
    - Windowed Standard Deviation

- o Data Interpreted as Stochastic Series
  - – Pearson product-moment correlation coefficient
  - – Spearman's rank correlation coefficient
- o Data Interpreted as a Probability Distribution Function
  - – Kolmogorov–Smirnov test
  - – Cramér–von Mises criterion

Visualization

Time series can be visualized with two categories of chart:Overlapping Charts and Separated Charts. Overlapping Charts display all time series on the same layout while Separated Charts presents them on different layouts (but aligned for comparison purpose)

Overlapping Charts

- Braided Graphs
- Line Charts
- Slope Graphs
- GapChart

Separated Charts

- Horizon Graphs
- Reduced Line Charts (small multiples)
- Silhouette Graph
- Circular Silhouette Graph

Time Series Regression and Exploratory Data Analysis

Time series regression models attempt to predict a future response based on the response history (known as autoregressive dynamics) and the transfer of dynamics from relevant predictors. Time series regression can help you understand and predict the behaviour of dynamic systems from experimental or observational data. Time series regression is commonly used for modelling and forecasting of economic, financial, and biological systems.

Exploratory Data Analysis

In statistics, exploratory data analysis (EDA) is an approach to analyzing data sets to summarize their main characteristics, often with visual methods. A statistical model can be used or not, but primarily EDA is for seeing what the data can tell us beyond the formal modelling or hypothesis testing task. Exploratory data analysis was promoted by

John Tukey to encourage statisticians to explore the data, and possibly formulate hypotheses that could lead to new data collection and experiments. EDA is different from initial data analysis (IDA), which focuses more narrowly on checking assumptions required for model fitting and hypothesis testing, and handling missing values and making transformations of variables as needed. EDA encompasses IDA.

Tukey's championing of EDA encouraged the development of statistical computing packages, especially S at Bell Labs. The S programming language inspired the systems 'S'-PLUS and R. This family of statistical-computing environments featured vastly improved dynamic visualization capabilities, which allowed statisticians to identify outliers, trends and patterns in data that merited further study.

Tukey's EDA was related to two other developments in statistical theory: Robust statistics and nonparametric statistics, both of which tried to reduce the sensitivity of statistical inferences to errors in formulating statistical models. Tukey promoted the use of five number summary of numerical data—the two extremes (maximum and minimum), the median, and the quartiles—because these median and quartiles, being functions of the empirical distribution are defined for all distributions, unlike the mean and standard deviation; moreover, the quartiles and median are more robust to skewed or heavy-tailed distributions than traditional summaries (the mean and standard deviation). The packages S, S-PLUS, and R included routines using resampling statistics, such as Quenouille and Tukey's jackknife and Efron's bootstrap, which are nonparametric and robust (for many problems).

Exploratory data analysis, robust statistics, nonparametric statistics, and the development of statistical programming languages facilitated statisticians' work on scientific and engineering problems. Such problems included the fabrication of semiconductors and the understanding of communications networks, which concerned Bell Labs. These statistical developments, all championed by Tukey, were designed to complement the analytic theory of testing statistical hypotheses, particularly the Laplacian tradition's emphasis on exponential families.

EDA Development

John W. Tukey wrote the book "Exploratory Data Analysis" in 1977. Tukey held that too much emphasis in statistics was placed on statistical hypothesis testing (confirmatory data analysis); more emphasis needed to be placed on using data to suggest hypotheses to test. In particular, he held that confusing the two types of analyses and employing them on the same set of data can lead to systematic bias owing to the issues inherent in testing hypotheses suggested by the data.

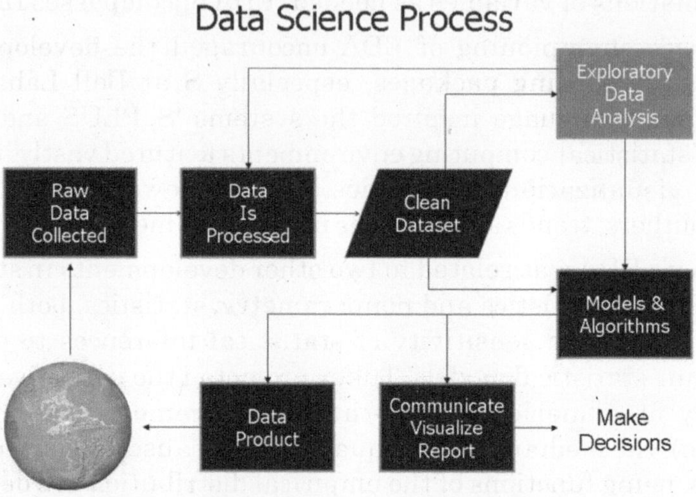

Figure: *Data science process flowchart*

The objectives of EDA are to:

- Suggest hypotheses about the causes of observed phenomena
- Assess assumptions on which statistical inference will be based
- Support the selection of appropriate statistical tools and techniques
- Provide a basis for further data collection through surveys or experiments

Many EDA techniques have been adopted into data mining and are being taught to young students as a way to introduce them to statistical thinking.

Techniques

There are a number of tools that are useful for EDA, but EDA is characterized more by the attitude taken than by particular techniques.

Typical graphical techniques used in EDA are:

Box Plot: In descriptive statistics, a box plot or boxplot is a convenient way of graphically depicting groups of numerical data through their quartiles. Box plots may also have lines extending vertically from the boxes (whiskers) indicating variability outside the upper and lower quartiles, hence the terms box-and-whisker plot and box-and-whisker diagram. Outliers may be plotted as individual points.

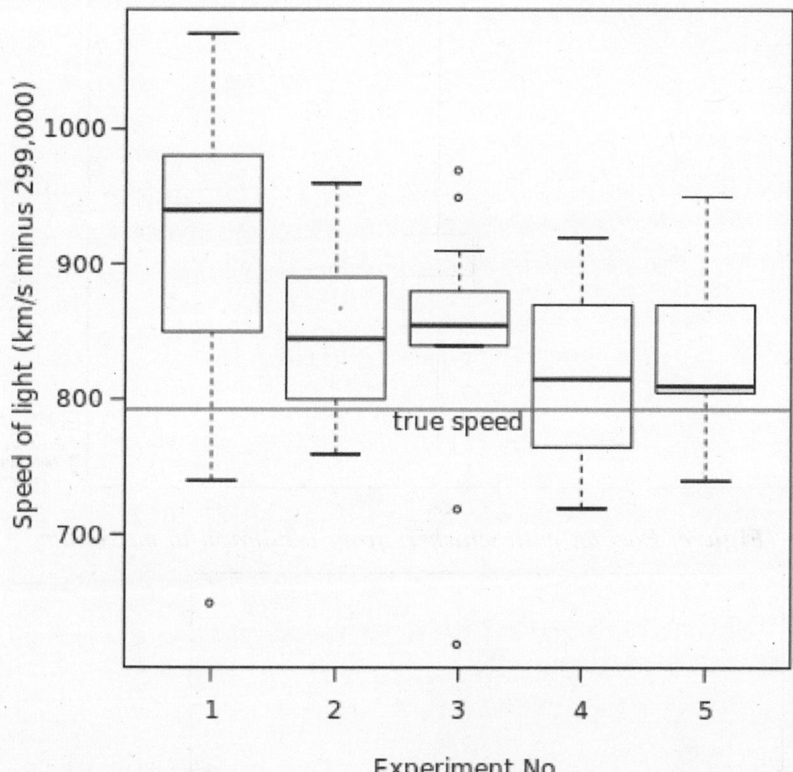

Figure: *Box plot of data from the Michelson–Morley experiment*

Box plots display variation in samples of a statistical population without making any assumptions of the underlying statistical distribution: box plots are non-parametric.

The spacings between the different parts of the box indicate the degree of dispersion (spread) and skewness in the data, and show outliers. In addition to the points themselves, they allow one to visually estimate various L-estimators, notably the interquartile range, midhinge, range, mid-range, and trimean. Boxplots can be drawn either horizontally or vertically.

Types of Boxplots

Box and whisker plots are uniform in their use of the box: the bottom and top of the box are always the first and third quartiles, and the band inside the box is always the second quartile (the median).

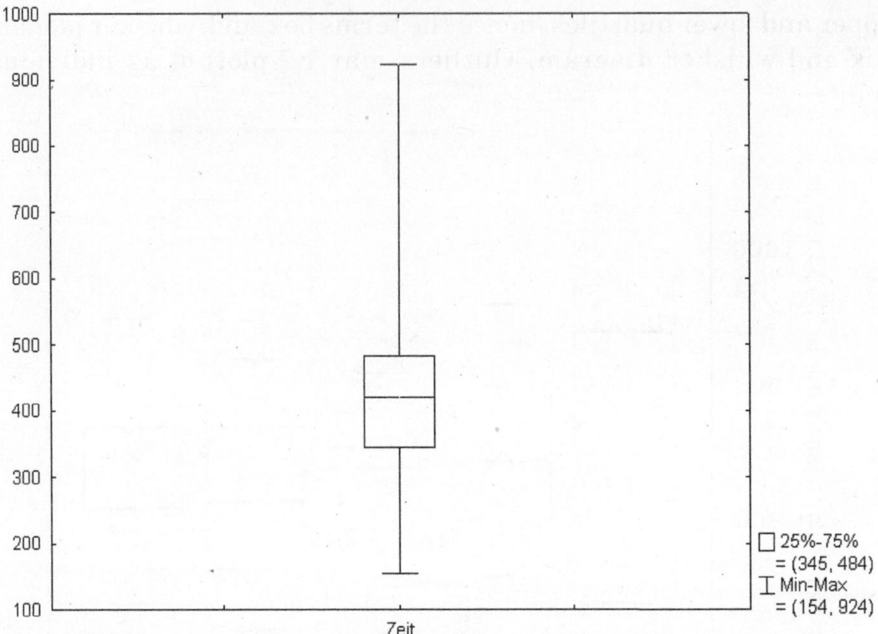

Figure: *Boxplot with whiskers from minimum to maximum*

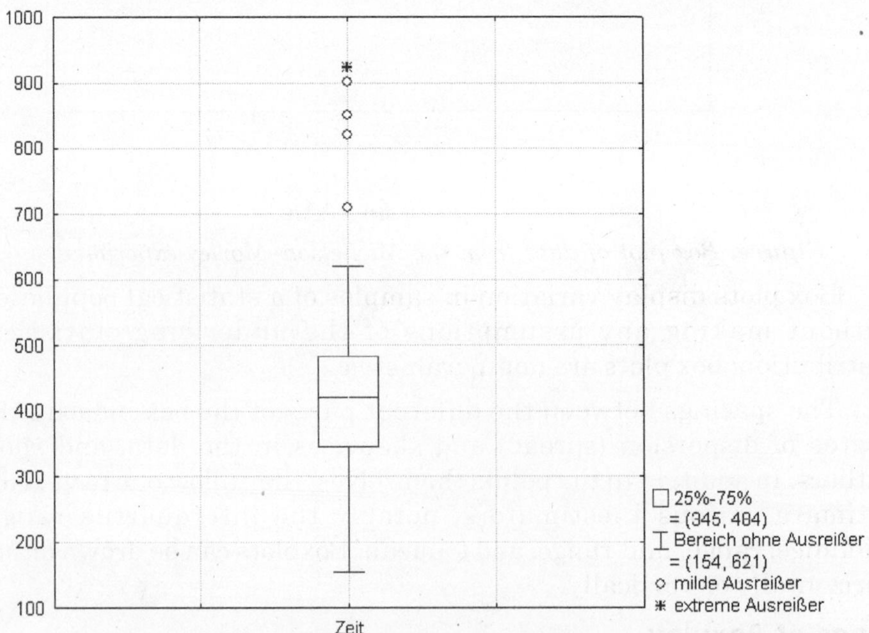

Figure: *Same Boxplot with whiskers with maximum 1.5 IQR*

But the ends of the whiskers can represent several possible alternative values, among them:

- the minimum and maximum of all of the data
- the lowest datum still within 1.5 IQR of the lower quartile, and the highest datum still within 1.5 IQR of the upper quartile (often called the Tukey boxplot)
- one standard deviation above and below the mean of the data
- the 9th percentile and the 91st percentile
- the 2nd percentile and the 98th percentile.

Any data not included between the whiskers should be plotted as an outlier with a dot, small circle, or star, but occasionally this is not done.

Some box plots include an additional character to represent the mean of the data.

On some box plots a crosshatch is placed on each whisker, before the end of the whisker.

Rarely, box plots can be presented with no whiskers at all.

Because of this variability, it is appropriate to describe the convention being used for the whiskers and outliers in the caption for the plot.

The unusual percentiles 2%, 9%, 91%, 98% are sometimes used for whisker cross-hatches and whisker ends to show the seven-number summary. If the data is normally distributed, the locations of the seven marks on the box plot will be equally spaced.

Variations

Since the American mathematician John W. Tukey introduced this type of visual data display in 1969, several variations on the traditional box plot have been described. Two of the most common are variable width box plots and notched box plots.

Variable width box plots illustrate the size of each group whose data is being plotted by making the width of the box proportional to the size of the group. A popular convention is to make the box width proportional to the square root of the size of the group.

Notched box plots apply a "notch" or narrowing of the box around the median. Notches are useful in offering a rough guide to significance of difference of medians; if the notches of two boxes do not overlap, this offers evidence of a statistically significant difference between the medians. The width of the notches is proportional to the interquartile range of the sample and inversely proportional to the square root of the size of the sample. However, there is uncertainty about the most appropriate multiplier (as this may vary depending on the similarity of the variances of the samples). One convention is to use $+/-1.58 * IQR / sqrt(n)$.

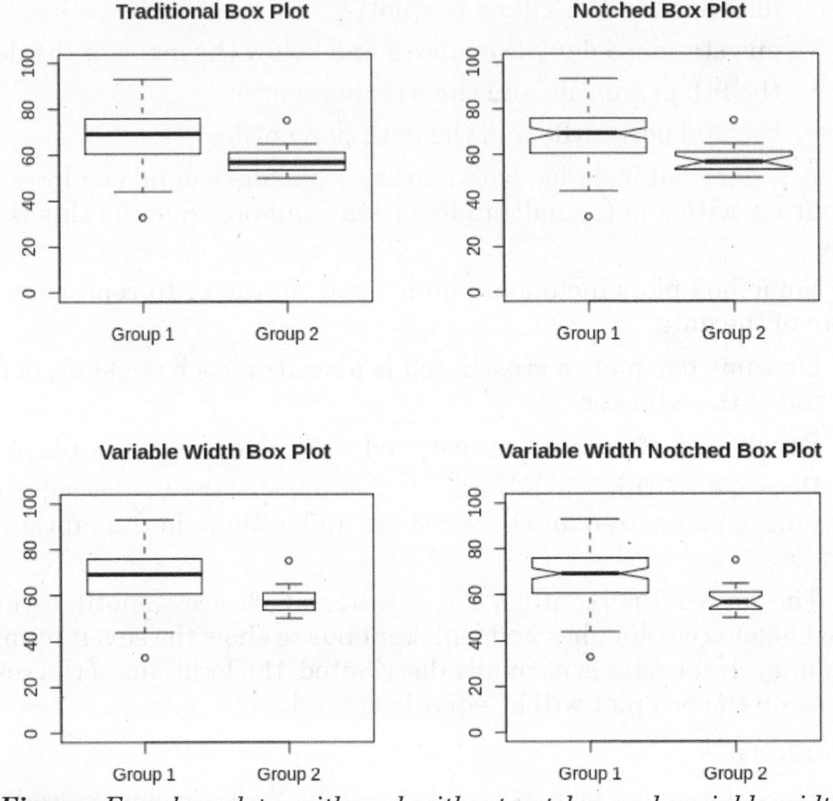

***Figure:** Four box plots, with and without notches and variable width*

Visualization

The box plot is a quick way of examining one or more sets of data graphically. Box plots may seem more primitive than a histogram or kernel density estimate but they do have some advantages. They take up less space and are therefore particularly useful for comparing distributions between several groups or sets of data. Choice of number and width of bins techniques can heavily influence the appearance of a histogram, and choice of bandwidth can heavily influence the appearance of a kernel density estimate.

As looking at a statistical distribution is more intuitive than looking at a box plot, comparing the box plot against the probability density function (theoretical histogram) for a normal $N(0,1\sigma^2)$ distribution may be a useful tool for understanding the box plot.

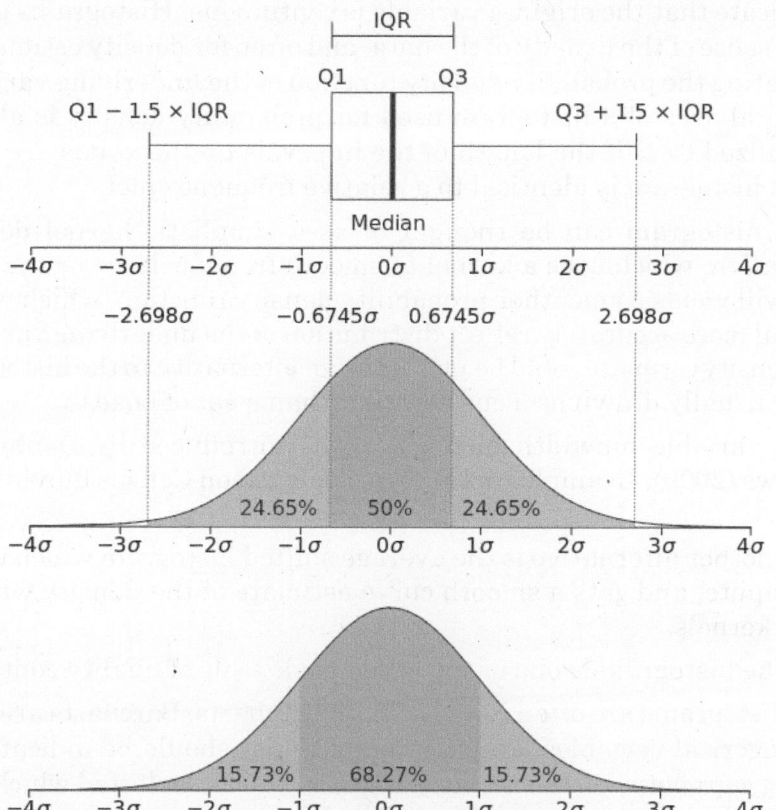

Figure: *Boxplot and a probability density function (pdf) of a Normal N(0,1s²) Population*

Histogram

A histogram is a graphical representation of the distribution of data. It is an estimate of the probability distribution of a continuous variable (quantitative variable) and was first introduced by Karl Pearson. To construct a histogram, the first step is to bin the range of values, and then count how many values fall into each interval. A rectangle is drawn with height proportional to the count and width equal to the bin size, so that rectangles abut each other.

A histogram may also be normalized displaying relative frequencies. It then shows the proportion of cases that fall into each of several categories, with the sum of the heights equaling 1. The bins are usually specified as consecutive, non-overlapping intervals of a variable. The bins (intervals) must be adjacent, and usually equal size. The rectangles of a histogram are drawn so that they touch each other to indicate that the original variable is continuous. Histograms give a rough sense of the density of the data, and often for density estimation: estimating the probability density function of the underlying variable. The total area of a histogram used for probability density is always normalized to 1. If the length of the intervals on the x-axis are all 1, then a histogram is identical to a relative frequency plot.

A histogram can be thought of as a simplistic kernel density estimation, which uses a kernel to smooth frequencies over the bins. This will yields a smoother probability density function, which will in general more accurately reflect distribution of the underlying variable. The density estimate could be plotted as an alternative to the histogram, and is usually drawn as a curve rather than a set of boxes.

A variable binwidth histogram was introduced by Denby and Mallows (2009). Examples of this are displayed on Census bureau data below.

Another alternative is the average shifted histogram which is fast to compute, and gets a smooth curve estimate of the density without using kernels.

The histogram is one of the seven basic tools of quality control.

Histograms are often confused with barcharts. Barcharts are plots of categorical variables, and the discontinuity should be indicated by having gaps between the rectangles. Often this is neglected which may lead to a barchart being confused for a histogram.

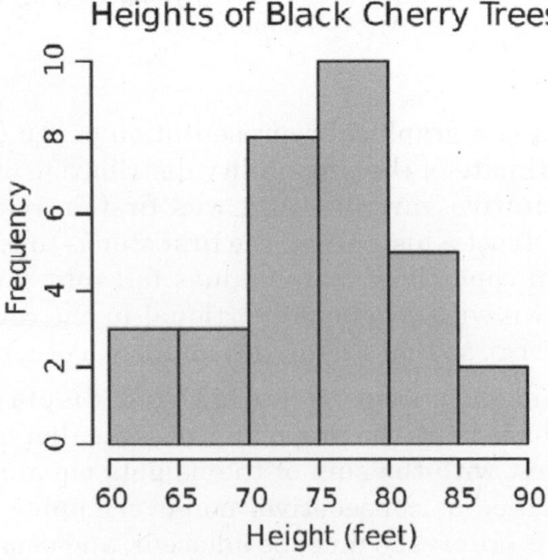

Figure: *An example histogram of the heights of 31 Black Cherry trees.*

Examples

This is a toy example

Bin	Count
-3.5	9
-2.5	32
-1.5	109
-0.5	180
0.5	132
1.5	34
2.5	4
3.5	9

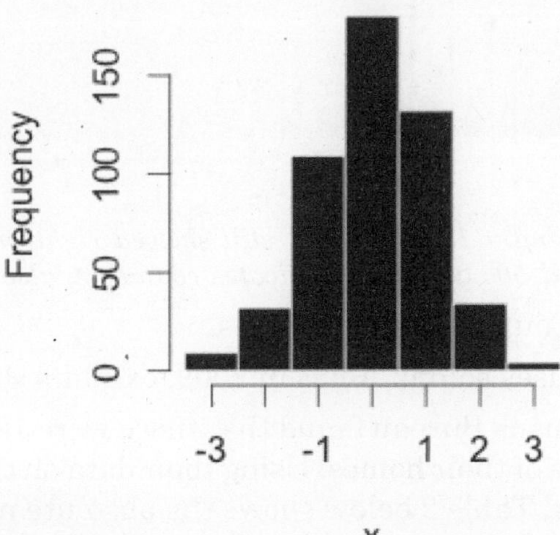

The language used to describe the patterns in a histogram are symmetric, skewed left or right, unimodal, bimodal or multimodal.

It is a good idea to plot your data on several different binwidths to learn more about it. Here is an example on tips given in a restaurant.

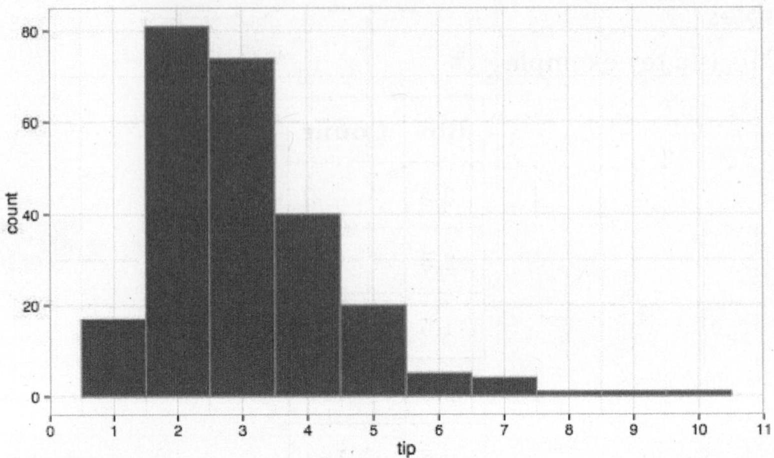

Figure: *Tips using a $1 binwidth, skewed right, unimodal*

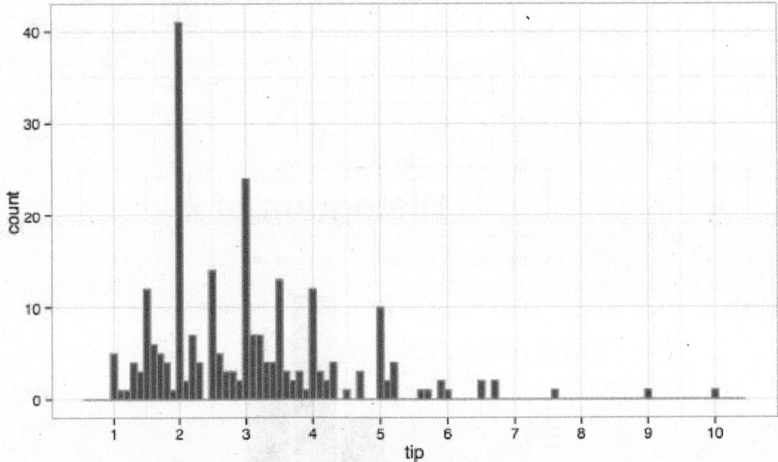

Figure: *Tips using a 10c binwidth, still skewed right, multimodal with modes at $ and 50c amounts, indicates rounding, also some outliers*

Here are a Couple More Examples.

Prices of houses sold in Ames in 2009, exhibits some right-skew.

The U.S. Census Bureau found that there were 124 million people who work outside of their homes. Using their data on the time occupied by travel to work, Table 2 below shows the absolute number of people who responded with travel times "at least 30 but less than 35 minutes" is higher than the numbers for the categories above and below it. This is likely due to people rounding their reported journey time. The problem of reporting values as somewhat arbitrarily rounded numbers is a common phenomenon when collecting data from people.

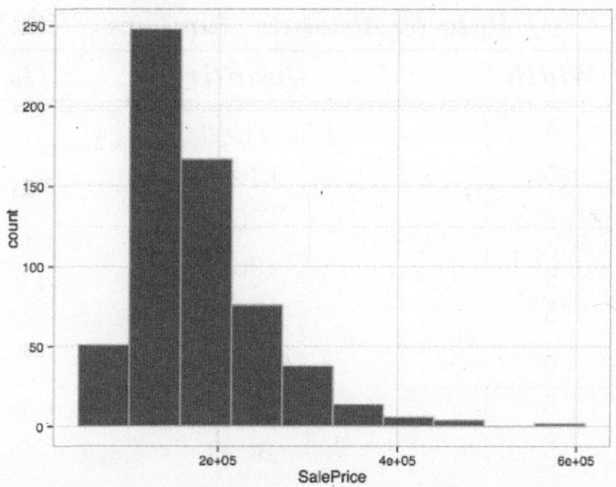

Figure: *Aces by players in a grand slam tennis tournament, facetted by gender. There are more aces in the mens game.*

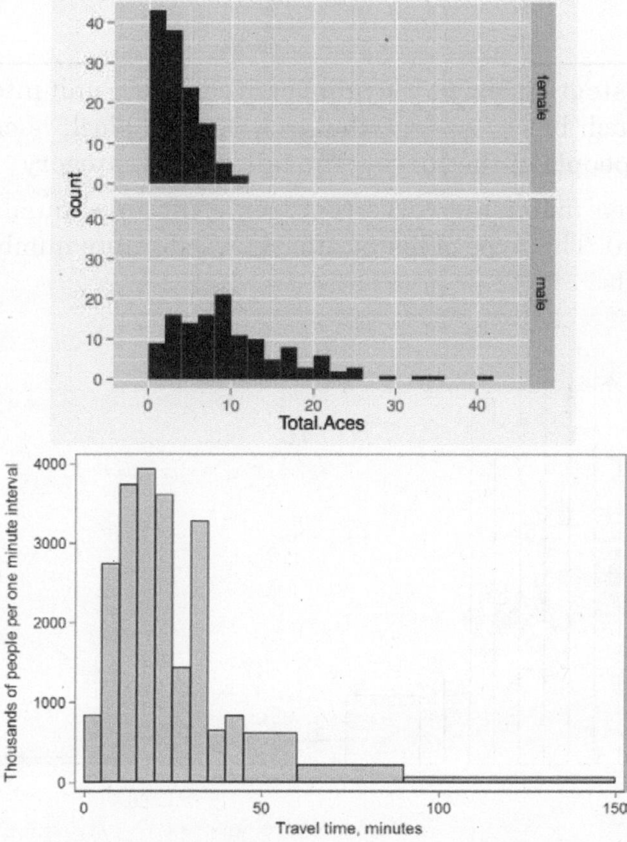

Figure: *Histogram of travel time (to work), US 2000 census. Area under the curve equals 1. This diagram uses Q/total/width from the table.*

Data by Absolute Numbers			
Interval	***Width***	***Quantity***	***Quantity/Width***
0	5	4180	836
5	5	13687	2737
10	5	18618	3723
15	5	19634	3926
20	5	17981	3596
25	5	7190	1438
30	5	16369	3273
35	5	3212	642
40	5	4122	824
45	15	9200	613
60	30	6461	215
90	60	3435	57

This histogram shows the number of cases per unit interval as the height of each block, so that the area of each block is equal to the number of people in the survey who fall into its category.

The area under the curve represents the total number of cases (124 million). This type of histogram shows absolute numbers, with Q in thousands.

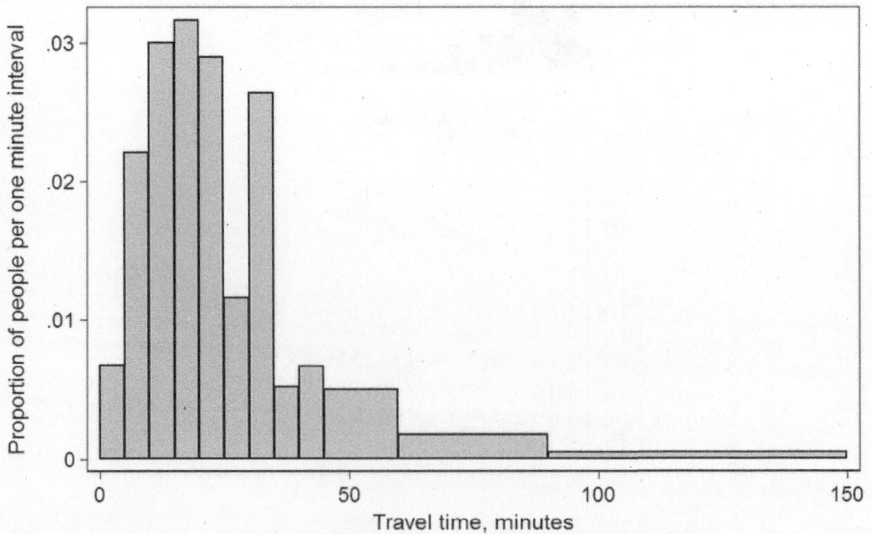

Figure: *Histogram of travel time (to work), US 2000 census. Area under the curve equals 1. This diagram uses Q/total/width from the table.*

Data by Proportion			
Interval	***Width***	***Quantity (Q)***	***Q/total/width***
0	5	4180	0.0067
5	5	13687	0.0221
10	5	18618	0.0300
15	5	19634	0.0316
20	5	17981	0.0290
25	5	7190	0.0116
30	5	16369	0.0264
35	5	3212	0.0052
40	5	4122	0.0066
45	15	9200	0.0049
60	30	6461	0.0017
90	60	3435	0.0005

This histogram differs from the first only in the vertical scale. The area of each block is the fraction of the total that each category represents, and the total area of all the bars is equal to 1 (the fraction meaning "all"). The curve displayed is a simple density estimate. This version shows proportions, and is also known as a unit area histogram.

In other words, a histogram represents a frequency distribution by means of rectangles whose widths represent class intervals and whose areas are proportional to the corresponding frequencies: the height of each is the average frequency density for the interval. The intervals are placed together in order to show that the data represented by the histogram, while exclusive, is also contiguous. (E.g., in a histogram it is possible to have two connecting intervals of 10.5–20.5 and 20.5–33.5, but not two connecting intervals of 10.5–20.5 and 22.5–32.5. Empty intervals are represented as empty and not skipped.)

Mathematical Definition

In a more general mathematical sense, a histogram is a function m_i that counts the number of observations that fall into each of the disjoint categories (known as bins), whereas the graph of a histogram is merely one way to represent a histogram. Thus, if we let n be the total number of observations and k be the total number of bins, the histogram m_i meets the following conditions:

$$n = \sum_{i=1}^{k} m_i.$$

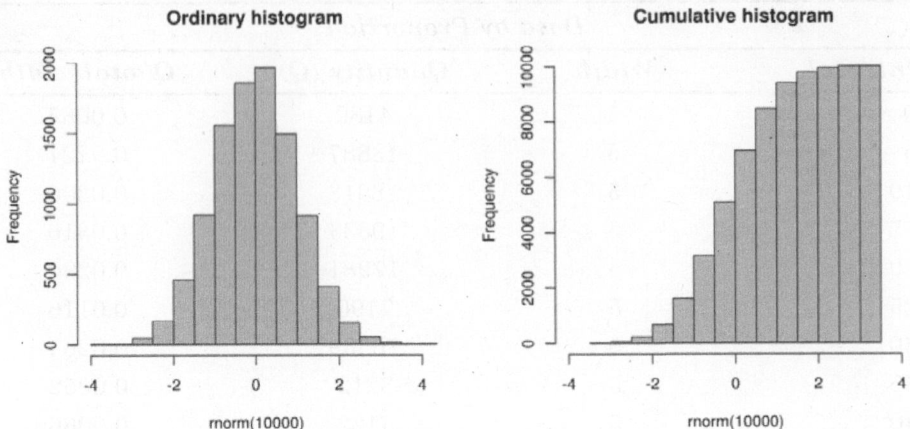

***Figure:** An ordinary and a cumulative histogram of the same data. The data shown is a random sample of 10,000 points from a normal distribution with a mean of 0 and a standard deviation of 1.*

Cumulative Histogram

A cumulative histogram is a mapping that counts the cumulative number of observations in all of the bins up to the specified bin. That is, the cumulative histogram M_i of a histogram m_j is defined as:

$$M_i = \sum_{j=1}^{i} m_j.$$

Number of Bins and Width

There is no "best" number of bins, and different bin sizes can reveal different features of the data. Grouping data is at least as old as Graunt's work in the 17th century, but no systematic guidelines were given until Sturges's work in 1926.

Using wider bins where the density is low reduces noise due to sampling randomness; using narrower bins where the density is high (so the signal drowns the noise) gives greater precision to the density estimation. Thus varying the bin-width within a histogram can be beneficial. Nonetheless, equal-width bins are widely used.

Some theoreticians have attempted to determine an optimal number of bins, but these methods generally make strong assumptions about the shape of the distribution. Depending on the actual data distribution and the goals of the analysis, different bin widths may be appropriate, so experimentation is usually needed to determine an appropriate width. There are, however, various useful guidelines and rules of thumb.

The number of bins k can be assigned directly or can be calculated from a suggested bin width h as:

$$k = \left\lceil \frac{\max x - \min x}{h} \right\rceil.$$

The braces indicate the ceiling function.

Square-root choice

$$k = \sqrt{n},$$

which takes the square root of the number of data points in the sample (used by Excel histograms and many others).

Sturges' Formula

Sturges' formula is derived from a binomial distribution and implicitly assumes an approximately normal distribution.

$$k = \lfloor \log_2 n + 1 \rfloor,$$

It implicitly bases the bin sizes on the range of the data and can perform poorly if n < 30, because the number of bins will be small—less than seven—and unlikely to show trends in the data well. It may also perform poorly if the data are not normally distributed.

Rice Rule

$$k = \lceil 2n^{1/3} \rceil,$$

The Rice Rule is presented as a simple alternative to Sturges's rule.

Doane's Formula

Doane's formula is a modification of Sturges' formula which attempts to improve its performance with non-normal data.

$$k = 1 + \log_2(n) + \log_2\left(1 + \frac{|g_1|}{\sigma_{g_1}}\right)$$

where g_1 is the estimated 3rd-moment-skewness of the distribution and

$$\sigma_{g_1} = \sqrt{\frac{6(n-2)}{(n+1)(n+3)}}$$

Scott's Normal Reference Rule

$$h = \frac{3.5\hat{\sigma}}{n^{1/3}},$$

where $\hat{\sigma}$ is the sample standard deviation. Scott's normal reference rule is optimal for random samples of normally distributed data, in the sense that it minimizes the integrated mean squared error of the density estimate.

Freedman–Diaconis' Choice

The Freedman–Diaconis rule is:

$$h = 2\frac{\text{IQR}(x)}{n^{1/3}},$$

which is based on the interquartile range, denoted by IQR. It replaces 3.5σ of Scott's rule with 2 IQR, which is less sensitive than the standard deviation to outliers in data.

Choice based on Minimization of an Estimated L^2 Risk Function

$$\underset{h}{\operatorname{argmin}} \frac{2\bar{m} - v}{h^2}$$

where $\bar{m}$ and v are mean and biased variance of a histogram with bin-width h, $\bar{m} = \frac{1}{k}\sum_{i=1}^{k} m_i$ and $v = \frac{1}{k}\sum_{i=1}^{k} (m_i - \bar{m})^2$.

Multi-Vari Chart

In quality control, multi-vari charts are a visual way of presenting variability through a series of charts. The content and format of the charts has evolved over time.

Original Concept

Multi-vari charts were first described by Leonard Seder in 1950, though they were developed independently by multiple sources. They were inspired by the stock market candlestick charts or open-high-low-close charts.

As originally conceived, the multi-vari chart resembles a Shewhart individuals control chart with the following differences:

- The quality characteristic of interest is measured at two extremes (around its diameter, along its length, or across its surface) and these measurements are plotted as vertical lines connecting the minimum and maximum values over time.

- The quality characteristic of interest is plotted across three horizontal panels that represent:
- Variability on a single piece
- Piece-to-piece variability
- Time-to-time variability
- The quality characteristic of interest is plotted against upper and lower specifications rather than control limits.

Recent Usage

More recently, the term "multi-vari chart" has been used to describe a visual way to display analysis of variance data (typically be expressed in tabular format). It consists of a series of panels which portray minimum, mean, and maximum responses for each treatment combination of interest rather than for periods of time.

Because it is a two-dimensional representation of multiple dimensions (one for each factor in the ANOVA), the multi-vari chart is only useful for comparing the variability among at most four factors.

The chart consists of the following:

- One horizontal panel for each level of the outermost factor
- One cluster of points representing the minimum, mean, and maximum responses for the particular treatment combination, connected by lines for each level of the innermost factor
- In the case of four factors, vertical panels for each level of the next-innermost factor
- As with control charts, the vertical axis depicts the quality characteristic of interest (or experimental response)

Run Chart

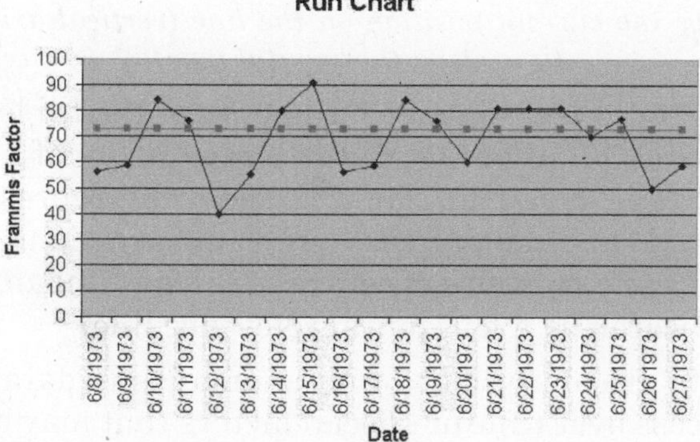

***Figure:** A simple run chart showing data collected over time. The median of the observed data (73) is also shown on the chart.*

A run chart, also known as a run-sequence plot is a graph that displays observed data in a time sequence. Often, the data displayed represent some aspect of the output or performance of a manufacturing or other business process.

Overview

Run sequence plots are an easy way to graphically summarize a univariate data set. A common assumption of univariate data sets is that they behave like:

- random drawings;
- from a fixed distribution;
- with a common location; and
- with a common scale.

With run sequence plots, shifts in location and scale are typically quite evident. Also, outliers can easily be detected.

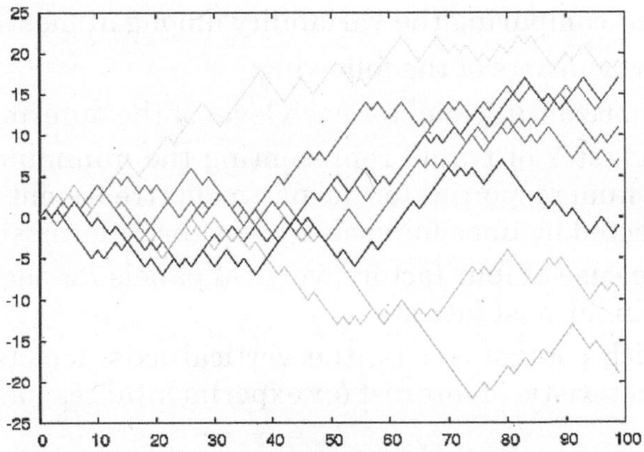

Figure: *Run chart of eight random walks in one dimension starting at 0. The plot shows the current position on the line (vertical axis) versus the time steps (horizontal axis).*

Examples could include measurements of the fill level of bottles filled at a bottling plant or the water temperature of a dishwashing machine each time it is run. Time is generally represented on the horizontal (x) axis and the property under observation on the vertical (y) axis. Often, some measure of central tendency (mean or median) of the data is indicated by a horizontal reference line.

Run charts are analyzed to find anomalies in data that suggest shifts in a process over time or special factors that may be influencing the variability of a process. Typical factors considered include unusually long "runs" of data points above or below the average line, the total number of such runs in the data set, and unusually long series of consecutive increases or decreases.

Run charts are similar in some regards to the control charts used in statistical process control, but do not show the control limits of the process. They are therefore simpler to produce, but do not allow for the full range of analytic techniques supported by control charts.

Pareto Chart

A Pareto chart, named after Vilfredo Pareto, is a type of chart that contains both bars and a line graph, where individual values are represented in descending order by bars, and the cumulative total is represented by the line.

The left vertical axis is the frequency of occurrence, but it can alternatively represent cost or another important unit of measure. The right vertical axis is the cumulative percentage of the total number of occurrences, total cost, or total of the particular unit of measure. Because the reasons are in decreasing order, the cumulative function is a concave function. To take the example above, in order to lower the amount of late arrivals by 78%, it is sufficient to solve the first three issues.

The purpose of the Pareto chart is to highlight the most important among a (typically large) set of factors. In quality control, it often represents the most common sources of defects, the highest occurring type of defect, or the most frequent reasons for customer complaints, and so on. Wilkinson (2006) devised an algorithm for producing statistically based acceptance limits (similar to confidence intervals) for each bar in the Pareto chart.

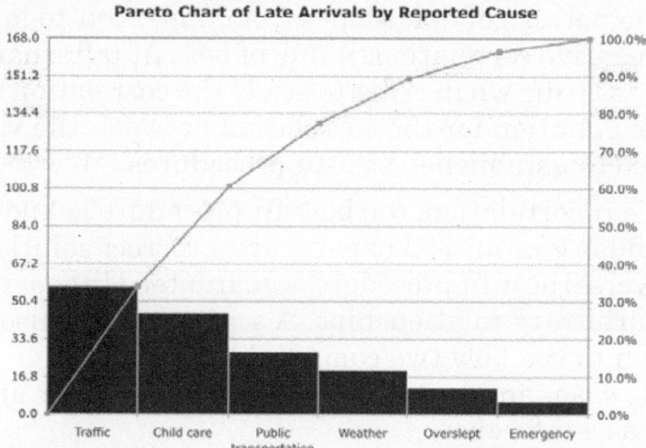

Figure: *Simple example of a Pareto chart using hypothetical data showing the relative frequency of reasons for arriving late at work*

These charts can be generated by simple spreadsheet programs, such as OpenOffice.org Calc and Microsoft Excel and specialized statistical software tools as well as online quality charts generators.

The Pareto chart is one of the seven basic tools of quality control.

Scatter Plot

A scatter plot, scatterplot, or scattergraph is a type of mathematical diagram using Cartesian coordinates to display values for two variables for a set of data. The data is displayed as a collection of points, each having the value of one variable determining the position on the horizontal axis and the value of the other variable determining the position on the vertical axis. This kind of plot is also called a scatter chart, scattergram, scatter diagram, or scatter graph.

A scatter plot is used when a variable exists that is below the control of the experimenter. If a parameter exists that is systematically incremented and/or decremented by the other, it is called the control parameter or independent variable and is customarily plotted along the horizontal axis. The measured or dependent variable is customarily plotted along the vertical axis. If no dependent variable exists, either type of variable can be plotted on either axis and a scatter plot will illustrate only the degree of correlation (not causation) between two variables.

A scatter plot can suggest various kinds of correlations between variables with a certain confidence interval. For example, weight and height, weight would be on x axis and height would be on the y axis. Correlations may be positive (rising), negative (falling), or null (uncorrelated). If the pattern of dots slopes from lower left to upper right, it suggests a positive correlation between the variables being studied. If the pattern of dots slopes from upper left to lower right, it suggests a negative correlation. A line of best fit (alternatively called 'trendline') can be drawn in order to study the correlation between the variables. An equation for the correlation between the variables can be determined by established best-fit procedures.

For a linear correlation, the best-fit procedure is known as linear regression and is guaranteed to generate a correct solution in a finite time. No universal best-fit procedure is guaranteed to generate a correct solution for arbitrary relationships. A scatter plot is also very useful when we wish to see how two comparable data sets agree with each other. In this case, an identity line, i.e., a y=x line, or an 1:1 line, is often drawn as a reference.

The more the two data sets agree, the more the scatters tend to concentrate in the vicinity of the identity line; if the two data sets are numerically identical, the scatters fall on the identity line exactly. One of the most powerful aspects of a scatter plot, however, is its ability to show nonlinear relationships between variables. Furthermore, if the data are represented by a mixture model of simple relationships, these relationships will be visually evident as superimposed patterns.

The scatter diagram is one of the seven basic tools of quality control.

Example

For example, to display a link between a person's lung capacity, and how long that person could hold his/her breath, a researcher would choose a group of people to study, then measure each one's lung capacity (first variable) and how long that person could hold his/her breath (second variable). The researcher would then plot the data in a scatter plot, assigning "lung capacity" to the horizontal axis, and "time holding breath" to the vertical axis.

A person with a lung capacity of 400 cl who held his/her breath for 21.7 seconds would be represented by a single dot on the scatter plot at the point (400, 21.7) in the Cartesian coordinates. The scatter plot of all the people in the study would enable the researcher to obtain a visual comparison of the two variables in the data set, and will help to determine what kind of relationship there might be between the two variables.

Stem-and-Leaf Plot

A stem-and-leaf display is a device for presenting quantitative data in a graphical format, similar to a histogram, to assist in visualizing the shape of a distribution. They evolved from Arthur Bowley's work in the early 1900s, and are useful tools in exploratory data analysis. Stemplots became more commonly used in the 1980s after the publication of John Tukey's book on exploratory data analysis in 1977. The popularity during those years is attributable to their use of monospaced (typewriter) typestyles that allowed computer technology of the time to easily produce the graphics. Modern computers' superior graphic capabilities have meant these techniques are less often used.

A stem-and-leaf display is often called a stemplot, but the latter term often refers to another chart type. A simple stem plot may refer to plotting a matrix of y values onto a common x axis, and identifying the common x value with a vertical line, and the individual y values with symbols on the line.

Unlike histograms, stem-and-leaf displays retain the original data to at least two significant digits, and put the data in order, thereby easing the move to order-based inference and non-parametric statistics. A basic stem-and-leaf display contains two columns separated by a vertical line. The left column contains the stems and the right column contains the leaves.

Constructing a Stem-and-Leaf Display

To construct a stem-and-leaf display, the observations must first be sorted in ascending order: this can be done most easily if working by hand by constructing a draft of the stem-and-leaf display with the leaves unsorted, then sorting the leaves to produce the final stem-and-leaf display. Here is the sorted set of data values that will be used in the following example:

44 46 47 49 63 64 66 68 68 72 72 75 76 81 84 88 106

Next, it must be determined what the stems will represent and what the leaves will represent. Typically, the leaf contains the last digit of the number and the stem contains all of the other digits. In the case of very large numbers, the data values may be rounded to a particular place value (such as the hundreds place) that will be used for the leaves. The remaining digits to the left of the rounded place value are used as the stem. In this example, the leaf represents the ones place and the stem will represent the rest of the number (tens place and higher).

The stem-and-leaf display is drawn with two columns separated by a vertical line. The stems are listed to the left of the vertical line. It is important that each stem is listed only once and that no numbers are skipped, even if it means that some stems have no leaves. The leaves are listed in increasing order in a row to the right of each stem. It is important to note that when there is a repeated number in the data (such as two 72s) then the plot must reflect such (so the plot would look like 7 | 2 2 5 6 when it has the numbers 72 72 75 76)

```
 4 | 4 6 7 9
 5 |
 6 | 3 4 6 8 8
 7 | 2 2 5 6
 8 | 1 4 8
 9 |
10 | 6
key: 6|3=63
leaf unit: 1.0
stem unit: 10.0
```

Rounding may be needed to create a stem-and-leaf display. Based on the following set of data, the stem plot below would be created:

-23.678758, -12.45, -3.4, 4.43, 5.5, 5.678, 16.87, 24.7, 56.8

For negative numbers, a negative is placed in front of the stem unit, which is still the value X / 10. Non-integers are rounded. This allowed the stem and leaf plot to retain its shape, even for more complicated data sets. As in this example below:

```
-2 | 4
-1 | 2
-0 | 3
 0 | 4 6 6
 1 | 7
 2 | 5
 3 |
 4 |
 5 | 7
```

key: -2|4=-24

Usage

Stem-and-leaf displays are useful for displaying the relative density and shape of the data, giving the reader a quick overview of distribution. They retain (most of) the raw numerical data, often with perfect integrity. They are also useful for highlighting outliers and finding the mode. However, stem-and-leaf displays are only useful for moderately sized data sets (around 15-150 data points). With very small data sets a stem-and-leaf displays can be of little use, as a reasonable number of data points are required to establish definitive distribution properties. A dot plot may be better suited for such data. With very large data sets, a stem-and-leaf display will become very cluttered, since each data point must be represented numerically. A box plot or histogram may become more appropriate as the data size increases.

The ease with which histograms can now be generated on computers has meant that stem-and-leaf displays are less used today than in the 1980s, when they first became widely utilized as a quick method of displaying information graphically by hand.

Parallel Coordinates

Parallel coordinates is a common way of visualizing high-dimensional geometry and analyzing multivariate data.

To show a set of points in an n-dimensional space, a backdrop is drawn consisting of n parallel lines, typically vertical and equally spaced. A point in n-dimensional space is represented as a polyline with vertices on the parallel axes; the position of the vertex on the ith axis corresponds to the ith coordinate of the point.

This visualization is closely related to time series visualization, except that it is applied to data where the axes do not correspond to points in time, and therefore do not have a natural order. Therefore, different axis arrangements may be of interest.

Higher Dimensions

Adding more dimensions in parallel coordinates (often abbreviated | |-coords or PCs) involves adding more axes. The value of parallel coordinates is that certain geometrical properties in high dimensions transform into easily seen 2D patterns. For example, a set of points on a line in n-space transforms to a set of polylines (or curves) in parallel coordinates all intersecting at n – 1 points. For n = 2 this yields a point-line duality pointing out why the mathematical foundations of parallel coordinates are developed in the Projective rather than Euclidean space.

Also known are the patterns corresponding to (hyper)planes, curves, several smooth (hyper)surfaces, proximities, convexity and recently non-orientability. Since the process maps a k-dimensional data onto a lower 2D space, some loss of information is expected. The loss of information can be measured using Parseval's identity (or energy norm).

Statistical Considerations

When used for statistical data visualisation there are three important considerations: the order, the rotation, and the scaling of the axes. The order of the axes is critical for finding features, and in typical data analysis many reorderings will need to be tried. Some authors have come up with ordering heuristics which may create illuminating orderings. The rotation of the axes is a translation in the parallel coordinates and if the lines intersected outside the parallel axes it can be translated between them by rotations. The simplest example of this is rotating the axis by 180 degrees. The necessity of scaling stems from the fact that the plot is based on interpolation (linear combination) of consecutive pairs of variables. Therefore, the variables must be in common scale, and there are many scaling methods to be considered as part of data preparation process that can reveal more informative views.

A smooth parallel coordinate plot is achieved with splines. In the smooth plot, every observation is mapped into a parametric line (or curve), which is smooth, continuous on the axes, and orthogonal to each parallel axis. This design emphasizes the quantization level for each data attribute. If one uses the Fourier interpolation of degree equals to the data dimensionality, then an Andrews plot is achieved.

Reading

Inselberg (Inselberg 1997) made a full review of how to visually read out parallel coords' relational patterns. When most lines between two parallel axis are somewhat parallel to each others, that suggests a positive relationship between these two dimensions. When lines cross in a kind of superposition of X-shapes, that's negative relationship. When lines cross randomly or are parallel, that show there is no particular relationship.

Limitations

In parallel coordinates, each axis can have at most two neighbouring axes (one on the left, and one on the right). For a d-dimensional data set, at most d-1 relationships can be shown at a time. In time series visualization, there exists a natural predecessor and successor; therefore in this special case, there exists a preferred arrangement. However when the axes do not have a unique order, finding a good axis arrangement requires the use of heuristics and experimentation. In order to explore more complex relationships, axes must be reordered.

By arranging the axes in 3-dimensional space (however, still in parallel, like nails in a nail bed), an axis can have more than two neighbours in a circle around the central attribute, and the arrangement problem gets easier (for example by using a minimum spanning tree). A prototype of this visualization is available as extension to the data mining software ELKI. However, the visualization is harder to interpret and interact with than a linear order.

Software

While there is a large amount of papers about parallel coordinates, there are only few notable software publicly available to convert databases into parallel coordinates graphics. Notable software are ELKI, GGobi, Macrofocus High-D, Mondrian, and ROOT. Libraries include Protovis.js, D3.js provide basic examples, while more complex examples are also available. D3.Parcoords.js (a D3-based library) and Macrofocus High-D API (a Java library) specifically dedicated to | | -coords graphic creation have also been published.

Odds Ratio

In statistics, the odds ratio (usually abbreviated "OR") is one of three main ways to quantify how strongly the presence or absence of property A is associated with the presence or absence of property B in a given population. If each individual in a population either does or does not have a property "A", (e.g. "high blood pressure"), and also either does or does not have a property "B" (e.g. "moderate alcohol consumption") where both properties are appropriately defined, then a ratio can be formed which quantitatively describes the association between the presence/absence of "A" (high blood pressure) and the presence/absence of "B" (moderate alcohol consumption) for individuals in the population. This ratio is the odds ratio (OR) and can be computed following these steps:

1. For a given individual that has "B" compute the odds that the same individual has "A"
2. For a given individual that does not have "B" compute the odds that the same individual has "A"
3. Divide the odds from step 1 by the odds from step 2 to obtain the odds ratio (OR).

The term "individual" in this usage does not have to refer to a human being, as a statistical population can measure any set of entities, whether living or inanimate.

If the OR is greater than 1, then having "A" is considered to be "associated" with having "B" in the sense that the having of "B" raises (relative to not-having "B") the odds of having "A". Note that this is not enough to establish that B is a contributing cause of "A": it could be that the association is due to a third property, "C", which is a contributing cause of both "A" and "B".

The two other major ways of quantifying association are the risk ratio ("RR") and the absolute risk reduction ("ARR"). In clinical studies and many other settings, the parameter of greatest interest is often actually the RR, which is determined in a way that is similar to the one just described for the OR, except using probabilities instead of odds. Frequently, however, the available data only allows the computation of the OR; notably, this is so in the case of case-control studies, as explained below. On the other hand, if one of the properties (say, A) is sufficiently rare (the "rare disease assumption"), then the OR of having A given that the individual has B is a good approximation to the corresponding RR (the specification "A given B" is needed because, while the OR treats the two properties symmetrically, the RR and other measures do not).

In a more technical language, the OR is a measure of effect size, describing the strength of association or non-independence between two binary data values. It is used as a descriptive statistic, and plays an important role in logistic regression.

Chapter 2

Multidimensional Scaling

Multidimensional scaling (MDS) is a means of visualizing the level of similarity of individual cases of a dataset. It refers to a set of related ordination techniques used in information visualization, in particular to display the information contained in a distance matrix. An MDS algorithm aims to place each object in N-dimensional space such that the between-object distances are preserved as well as possible. Each object is then assigned coordinates in each of the N dimensions. The number of dimensions of an MDS plot N can exceed 2 and is specified a priori. Choosing N=2 optimizes the object locations for a two-dimensional scatterplot.

Types

MDS algorithms fall into a taxonomy, depending on the meaning of the input matrix:

Classical multidimensional scaling

Also known as Principal Coordinates Analysis, Torgerson Scaling or Torgerson–Gower scaling. Takes an input matrix giving dissimilarities between pairs of items and outputs a coordinate matrix whose configuration minimizes a loss function called strain.

Metric multidimensional scaling

A superset of classical MDS that generalizes the optimization procedure to a variety of loss functions and input matrices of known distances with weights and so on. A useful loss function in this context is called stress, which is often minimized using a procedure called stress majorization.

Non-metric multidimensional scaling

In contrast to metric MDS, non-metric MDS finds both a non-parametric monotonic relationship between the dissimilarities in the item-item matrix and the Euclidean distances between items, and the location of each item in the low-dimensional space. The relationship is typically found using isotonic regression.

- Louis Guttman's smallest space analysis (SSA) is an example of a non-metric MDS procedure.

Generalized multidimensional scaling

An extension of metric multidimensional scaling, in which the target space is an arbitrary smooth non-Euclidean space. In cases where the dissimilarities are distances on a surface and the target space is another surface, GMDS allows finding the minimum-distortion embedding of one surface into another.

Details

The data to be analyzed is a collection of I objects (colours, faces, stocks, . . .) on which a distance function is defined,

$$\delta_{i,j} := \text{distance between } i\text{-th and } j\text{-th objects.}$$

These distances are the entries of the dissimilarity matrix

$$\Delta := \begin{pmatrix} \delta_{1,1} & \delta_{1,2} & \cdots & \delta_{1,I} \\ \delta_{2,1} & \delta_{2,2} & \cdots & \delta_{2,I} \\ \vdots & \vdots & & \vdots \\ \delta_{I,1} & \delta_{I,2} & \cdots & \delta_{I,I} \end{pmatrix}.$$

The goal of MDS is, given Δ, to find vectors $x_1,\ldots,x_I \in \mathbb{R}^N$ such that

$$\| x_i - x_j \| \approx \delta_{i,j} \text{ for all } i,j \in 1,\ldots,I,$$

where $\|\cdot\|$ is a vector norm. In classical MDS, this norm is the Euclidean distance, but, in a broader sense, it may be a metric or arbitrary distance function.

In other words, MDS attempts to find an embedding from the I objects into $\mathbb{R}^N$ such that distances are preserved. If the dimension N is chosen x_i to be 2 or 3, we may plot the vectors x_i to obtain a visualization of the similarities between the I objects. Note that the vectors x_i are not unique: With the Euclidean distance, they may be arbitrarily translated, rotated, and reflected, since these transformations do not change the pairwise distances $\| x_i - x_j \|$.

(Note: The symbol $\mathbb{R}$ indicates the set of real numbers, and the notation $\mathbb{R}^N$ refers to the Cartesian product of N copies of $\mathbb{R}$, which is an N-dimensional vector space over the field of the real numbers.)

There are various approaches to determining the vectors x_i. Usually, MDS is formulated as an optimization problem, where $(x_1,\ldots,x_I)$ is found as a minimizer of some cost function, for example,

$$\min_{x_1,\ldots,x_I} \sum_{i<j} (\|x_i - x_j\| - \delta_{i,j})^2.$$

A solution may then be found by numerical optimization techniques. For some particularly chosen cost functions, minimizers can be stated analytically in terms of matrix eigendecompositions.

Procedure

There are several steps in conducting MDS research:

1. Formulating the problem – What variables do you want to compare? How many variables do you want to compare? More than 20 is often considered cumbersome. Fewer than 8 (4 pairs) will not give valid results. What purpose is the study to be used for?
2. Obtaining input data – Respondents are asked a series of questions. For each product pair, they are asked to rate similarity (usually on a 7 point Likert scale from very similar to very dissimilar). The first question could be for Coke/Pepsi for example, the next for Coke/Hires rootbeer, the next for Pepsi/ Dr Pepper, the next for Dr Pepper/Hires rootbeer, etc. The number of questions is a function of the number of brands and can be calculated as $Q = N(N-1)/2$ where Q is the number of questions and N is the number of brands. This approach is referred to as the "Perception data : direct approach". There are two other approaches. There is the "Perception data : derived approach" in which products are decomposed into attributes that are rated on a semantic differential scale. The other is the "Preference data approach" in which respondents are asked their preference rather than similarity.
3. Running the MDS statistical program – Software for running the procedure is available in many software for statistics. Often there is a choice between Metric MDS (which deals with interval

or ratio level data), and Nonmetric MDS (which deals with ordinal data).

4. Decide number of dimensions – The researcher must decide on the number of dimensions they want the computer to create. The more dimensions, the better the statistical fit, but the more difficult it is to interpret the results.
5. Mapping the results and defining the dimensions – The statistical program (or a related module) will map the results. The map will plot each product (usually in two-dimensional space). The proximity of products to each other indicate either how similar they are or how preferred they are, depending on which approach was used. How the dimensions of the embedding actually correspond to dimensions of system behaviour, however, are not necessarily obvious. Here, a subjective judgment about the correspondence can be made.
6. Test the results for reliability and validity – Compute R-squared to determine what proportion of variance of the scaled data can be accounted for by the MDS procedure. An R-square of 0.6 is considered the minimum acceptable level. An R-square of 0.8 is considered good for metric scaling and .9 is considered good for non-metric scaling. Other possible tests are Kruskal's Stress, split data tests, data stability tests (i.e., eliminating one brand), and test-retest reliability.
7. Report the results comprehensively – Along with the mapping, at least distance measure (e.g., Sorenson index, Jaccard index) and reliability (e.g., stress value) should be given. It is also very advisable to give the algorithm (e.g., Kruskal, Mather), which is often defined by the program used (sometimes replacing the algorithm report), if you have given a start configuration or had a random choice, the number of runs, the assessment of dimensionality, the Monte Carlo method results, the number of iterations, the assessment of stability, and the proportional variance of each axis (r-square).

Applications

Applications include scientific visualisation and data mining in fields such as cognitive science, information science, psychophysics, psychometrics, marketing and ecology. New applications arise in the scope of autonomous wireless nodes that populate a space or an area. MDS may apply as a real time enhanced approach to monitoring and managing such populations.

Furthermore, MDS has been used extensively in geostatistics, for modelling the spatial variability of the patterns of an image (by representing them as points in a lower-dimensional space), and natural language processing, for modelling the semantic and affective relatedness of natural language concepts (by representing them as points in a 100-dimensional vector space).

Marketing

In marketing, MDS is a statistical technique for taking the preferences and perceptions of respondents and representing them on a visual grid, called perceptual maps. By mapping multiple attributes and multiple brands at the same time, a greater understanding of the marketplace and of consumers' perceptions can be achieved, as compared with a basic two attribute perceptual map.

Comparison and Advantages

Potential customers are asked to compare pairs of products and make judgments about their similarity. Whereas other techniques (such as factor analysis, discriminant analysis, and conjoint analysis) obtain underlying dimensions from responses to product attributes identified by the researcher, MDS obtains the underlying dimensions from respondents' judgments about the similarity of products. This is an important advantage. It does not depend on researchers' judgments. It does not require a list of attributes to be shown to the respondents. The underlying dimensions come from respondents' judgments about pairs of products. Because of these advantages, MDS is the most common technique used in perceptual mapping.

Targeted Projection Pursuit

Targeted projection pursuit is a type of statistical technique used for exploratory data analysis, information visualization, and feature selection. It allows the user to interactively explore very complex data (typically having tens to hundreds of attributes) to find features or patterns of potential interest.

Conventional, or 'blind', projection pursuit, finds the most "interesting" possible projections in multidimensional data, using a search algorithm that optimizes some fixed criterion of "interestingness" – such as deviation from a normal distribution. In contrast, targeted projection pursuit allows the user to explore the space of projections by manipulating data points directly in an interactive scatter plot.

Targeted projection pursuit has found applications in DNA microarray data analysis, protein sequence analysis, graph layout and digital signal processing. It is available as a package for the WEKA machine learning toolkit.

Multilinear PCA

Multilinear principal component analysis (MPCA) is a mathematical procedure that uses multiple orthogonal transformations to convert a set of multidimensional objects into another set of multidimensional objects of lower dimensions. There is one orthogonal (linear) transformation for each dimension (mode); hence multilinear. This transformation aims to capture as high a variance as possible, accounting for as much of the variability in the data as possible, subject to the constraint of mode-wise orthogonality.

MPCA is a multilinear extension of principal component analysis (PCA). The major difference is that PCA needs to reshape a multidimensional object into a vector, while MPCA operates directly on multidimensional objects through mode-wise processing. For example, for 100x100 images, PCA operates on vectors of 10000x1 while MPCA operates on vectors of 100x1 in two modes. For the same amount of dimension reduction, PCA needs to estimate 49*(10000/(100*2)-1) times more parameters than MPCA. Thus, MPCA is more efficient and better conditioned in practice.

MPCA is a basic algorithm for dimension reduction via multilinear subspace learning. In wider scope, it belongs to tensor-based computation. Its origin can be traced back to the Tucker decomposition in 1960s and it is closely related to higher-order singular value decomposition, (HOSVD) and to the best rank-(R1, R2, ..., RN) approximation of higher-order tensors.

MPCA performs feature extraction by determining a multilinear projection that captures most of the original tensorial input variations. As in PCA, MPCA works on centred data. The MPCA solution follows the alternating least square (ALS) approach. Thus, is iterative in nature and it proceeds by decomposing the original problem to a series of multiple projection subproblems. Each subproblem is a classical PCA problem, which can be easily solved.

It should be noted that while PCA with orthogonal transformations produces uncorrelated features/variables, this is not the case for MPCA. Due to the nature of tensor-to-tensor transformation, MPCA features are not uncorrelated in general although the transformation in each mode is orthogonal. In contrast, the uncorrelated MPCA (UMPCA) generates uncorrelated multilinear features.

Feature Selection

MPCA produces tensorial features. For conventional usage, vectorial features are often preferred. For example most classifiers in the literature takes vectors as input. On the other hand, as there are correlations among MPCA features, a further selection process often improves the performance. Supervised (discriminative) MPCA feature selection is used in object recognition while unsupervised MPCA feature selection is employed in visualization task.

Typical quantitative techniques are:

Median Polish

The median polish is an exploratory data analysis procedure proposed by the statistician John Tukey. It finds an additively-fit model for data in a two-way layout table (usually, results from a factorial experiment) of the form row effect + column effect + overall median.

Trimean

In statistics the trimean (TM), or Tukey's trimean, is a measure of a probability distribution's location defined as a weighted average of the distribution's median and its two quartiles:

$$TM = \frac{Q_1 + 2Q_2 + Q_3}{4}$$

This is equivalent to the average of the median and the midhinge:

$$TM = \frac{1}{2}\left(Q_2 + \frac{Q_1 + Q_3}{2}\right)$$

The foundations of the trimean were part of Arthur Bowley's teachings, and later popularized by statistician John Tukey in his 1977 book which has given its name to a set of techniques called Exploratory data analysis.

Like the median and the midhinge, but unlike the sample mean, it is a statistically resistant L-estimator with a breakdown point of 25%.

Despite its simplicity, the trimean is a remarkably efficient estimator of population mean. More precisely, for a large data set (over 100 points) from a symmetric population, the average of the 20th, 50th, and 80th percentile is the most efficient 3 point L-estimator, with 88% efficiency. For context, the best 1 point estimate by L-estimators is the median, with an efficiency of 64% or better (for all n), while using 2 points (for a large data set of over 100 points from a symmetric population), the most efficient estimate is the 29% midsummary (mean of 29th and 71st percentiles), which has an efficiency of about 81%. Using quartiles, these optimal estimators can be approximated by the midhinge and the trimean. Using further points yield higher efficiency, though it is notable that only 3 points are needed for very high efficiency.

Ordination

In multivariate analysis, ordination or gradient analysis is a method complementary to data clustering, and used mainly in exploratory data analysis (rather than in hypothesis testing). Ordination orders objects that are characterized by values on multiple variables (i.e., multivariate objects) so that similar objects are near each other and dissimilar objects are farther from each other. These relationships between the objects, on each of several axes (one for each variable), are then characterized numerically and/or graphically. Many ordination techniques exist, including principal components analysis (PCA), non-metric multidimensional scaling (NMDS), correspondence analysis (CA) and its derivatives (detrended CA (DCA), canonical CA (CCA)), Bray–Curtis ordination, and redundancy analysis (RDA), among others.

Ordination can be used on the analysis of any set of multivariate objects. It is frequently used in several environmental or ecological sciences, particularly plant community ecology. It is also used in genetics and systems biology for microarray data analysis, and in psychometrics.

History

Many EDA ideas can be traced back to earlier authors, for example:

- Francis Galton emphasized order statistics and quantiles.
- Arthur Lyon Bowley used precursors of the stemplot and five-number summary (Bowley actually used a "seven-figure summary", including the extremes, deciles and quartiles, along with the median).
- Andrew Ehrenberg articulated a philosophy of data reduction.

The Open University course Statistics in Society (MDST 242), took the above ideas and merged them with Gottfried Noether's work, which introduced statistical inference via coin-tossing and the median test.

Data Point

In statistics, a data point or observation is a set of one or more measurements on a single member of a statistical population. For example, in a study of the determinants of money demand with the unit of observation being the individual, a data point might be the values of income, wealth, age of individual, number of dependents. Statistical inference about the population would be conducted using a statistical sample consisting of various such data points.

In addition, in statistical graphics, a "data point" may be an individual item with a statistical display; such points may relate to either a single member of a population or to a summary statistic calculated for a given subpopulation.

Discussion

The measurements contained in a data point are formally typed, where here type is used in a way compatible with datatype in computing; so that the type of measurement can specify whether the measurement results in a Boolean value from {yes, no}, an integer or real number, the identity of some category, or some vector or array. The implication of point is often that the data may be plotted in a graphic display, but in many cases the data are processed numerically before that is done. In the context of statistical graphics, measured values for individuals or summary statistics for different subpopulations are displayed as separate symbols within a display; since such symbols can differ by shape, size and colour, a single data point within a display can convey multiple aspects of the set of measurements for an individual or subpopulation.

Line Chart

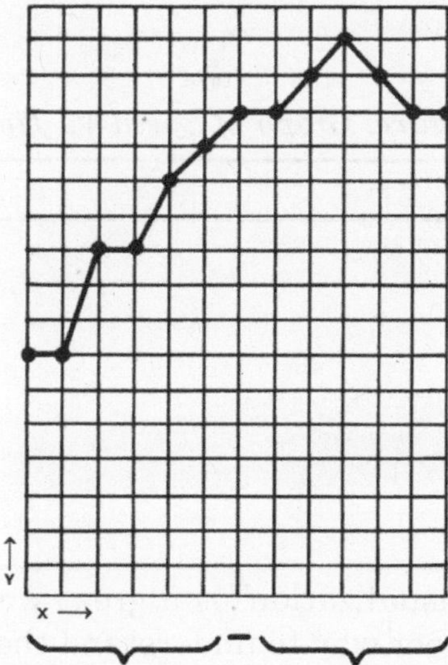

Figure: *This simple graph shows data over intervals with connected points*

A line chart or line graph is a type of chart which displays information as a series of data points called 'markers' connected by straight line segments. It is a basic type of chart common in many fields. It is similar to a scatter plot except that the measurement points are ordered (typically by their x-axis value) and joined with straight line segments. Line Charts show how a particular data changes at equal intervals of time. A line chart is often used to visualize a trend in data over intervals of time – a time series – thus the line is often drawn chronologically.

Example

In the experimental sciences, data collected from experiments are often visualized by a graph. For example, if one were to collect data on the speed of a body at certain points in time, one could visualize the data by a data table such as the following:

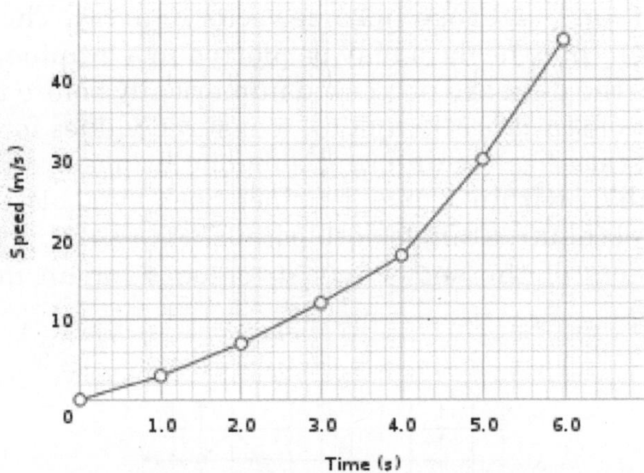

Figure: *Graph of Speed Vs Time*

Elapsed Time (s)	*Speed ($m\ s^{-1}$)*
0	0
1	3
2	7
3	12
4	20
5	30
6	45.6

Fig. The table "visualization" is a great way of displaying exact values, but can be a poor way to understand the underlying patterns that those values represent. Because of these qualities, the table display is often erroneously conflated with the data itself; whereas it is just another visualization of the data.

Understanding the process described by the data in the table is aided by producing a graph or line chart of Speed versus Time. Such a visualisation appears in the figure to the right.

Mathematically, if we denote time by the variable *t*, and speed by *v*, then the function plotted in the graph would be denoted *v(t)* indicating that *v* (the dependent variable) is a function of .

Best-Fit

Charts often include an overlaid mathematical function depicting the best-fit trend of the scattered data. This layer is referred to as a best-fit layer and the graph containing this layer is often referred to as a line graph. It is simple to construct a "best-fit" layer consisting of a set of line segments connecting adjacent data points; however, such a "best-fit" is usually not an ideal representation of the trend of the underlying scatter data for the following reasons:

1. It is highly improbable that the discontinuities in the slope of the best-fit would correspond exactly with the positions of the measurement values.
2. It is highly unlikely that the experimental error in the data is negligible, yet the curve falls exactly through each of the data points.

In either case, the best-fit layer can reveal trends in the data. Further, measurements such as the gradient or the area under the curve can be made visually, leading to more conclusions or results from the data.

A true best-fit layer should depict a continuous mathematical function whose parameters are determined by using a suitable error-minimization scheme, which appropriately weights the error in the data values. Such curve fitting functionality is often found in graphing software or spreadsheets. Best-fit curves may vary from simple linear equations to more complex quadratic, polynomial, exponential, and periodic curves.

Data Visualization

Data visualization or data visualisation is viewed by many disciplines as a modern equivalent of visual communication. It is not owned by any one field, but rather finds interpretation across many (e.g. it is viewed as a modern branch of descriptive statistics by some, but also as a grounded theory development tool by others). It involves the creation and study of the visual representation of data, meaning "information that has been abstracted in some schematic form, including attributes or variables for the units of information".

A primary goal of data visualization is to communicate information clearly and efficiently to users via the information graphics selected, such as tables and charts. Effective visualization helps users in analyzing and reasoning about data and evidence. It makes complex data more accessible, understandable and usable. Users may have particular analytical tasks, such as making comparisons or understanding causality, and the design principle of the graphic (i.e., showing comparisons or showing causality) follows the task. Tables are generally used where users will look-up a specific measure of a variable, while charts of various types are used to show patterns or relationships in the data for one or more variables.

Data visualization is both an art and a science. The rate at which data is generated has increased, driven by an increasingly information-based economy. Data created by internet activity and an expanding number of sensors in the environment, such as satellites and traffic cameras, are referred to as "Big Data". Processing, analyzing and communicating this data present a variety of ethical and analytical challenges for data visualization. The field of data science and practitioners called data scientists have emerged to help address this challenge.

According to Friedman (2008) the "main goal of data visualization is to communicate information clearly and effectively through graphical means. It doesn't mean that data visualization needs to look boring to be functional or extremely sophisticated to look beautiful. To convey ideas effectively, both aesthetic form and functionality need to go hand in hand, providing insights into a rather sparse and complex data set by communicating its key-aspects in a more intuitive way. Yet designers often fail to achieve a balance between form and function, creating gorgeous data visualizations which fail to serve their main purpose — to communicate information".

Indeed, Fernanda Viegas and Martin M. Wattenberg have suggested that an ideal visualization should not only communicate clearly, but stimulate viewer engagement and attention.

Data visualization is closely related to information graphics, information visualization, scientific visualization, exploratory data analysis and statistical graphics. In the new millennium, data visualization has become an active area of research, teaching and development. According to Post et al. (2002), it has united scientific and information visualization. Brian Willison has demonstrated that data visualization has also been linked to enhancing agile software development and customer engagement.

KPI Library has developed the "Periodic Table of Visualization Methods," an interactive chart displaying various data visualization methods. It includes six types of data visualization methods: data, information, concept, strategy, metaphor and compound.

In February 2014, University of Toronto professor Nadia Amoroso demonstrated how Data Visualization techniques increase the understanding of Big Data sets in order to communicate a story at the University of Waterloo Stratford Campus Inspiration Day.

Characteristics of Effective Graphical Displays

Professor Edward Tufte explained that users of information displays are executing particular analytical tasks such as making comparisons or determining causality. The design principle of the information graphic should support the analytical task, showing the comparison or causality.

In his 1983 book The Visual Display of Quantitative Information, Edward Tufte defines 'graphical displays' and principles for effective graphical display in the following passage: "Excellence in statistical graphics consists of complex ideas communicated with clarity, precision and efficiency. Graphical displays should:

- show the data
- induce the viewer to think about the substance rather than about methodology, graphic design, the technology of graphic production or something else
- avoid distorting what the data have to say
- present many numbers in a small space
- make large data sets coherent
- encourage the eye to compare different pieces of data
- reveal the data at several levels of detail, from a broad overview to the fine structure.
- serve a reasonably clear purpose: description, exploration, tabulation or decoration.
- be closely integrated with the statistical and verbal descriptions of a data set.

Graphics reveal data. Indeed graphics can be more precise and revealing than conventional statistical computations."

For example, the Minard diagram shows the losses suffered by Napoleon's army in the 1812-1813 period. Six variables are plotted: the size of the army, its location on a two-dimensional surface (x and y), time, direction of movement, and temperature. This multivariate display on a two dimensional surface tells a story that can be grasped immediately while identifying the source data to build credibility. Tufte wrote in 1983 that: "It may well be the best statistical graphic ever drawn."

Not applying these principles may result in misleading graphs, which distort the message or support an erroneous conclusion. According to Tufte, chartjunk refers to extraneous interior decoration of the graphic that does not enhance the message, or gratuitous three dimensional or perspective effects. Needlessly separating the explanatory key from the image itself, requiring the eye to travel back and forth from the image to the key, is a form of "administrative debris." The ratio of "data to ink" should be maximized, erasing non-data ink where feasible.

Quantitative Messages

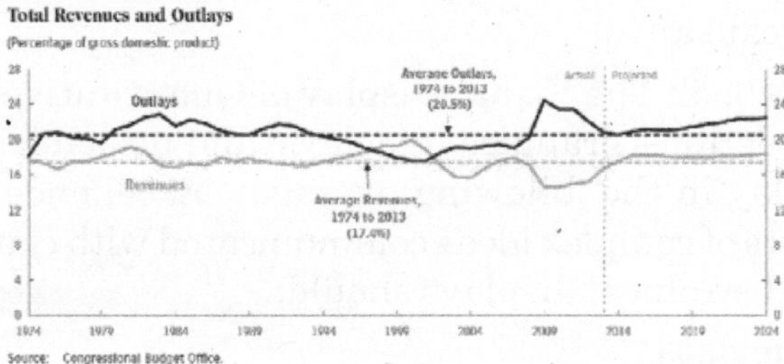

Figure: *A time series illustrated with a line chart demonstrating trends in U.S. federal spending and revenue over time.*

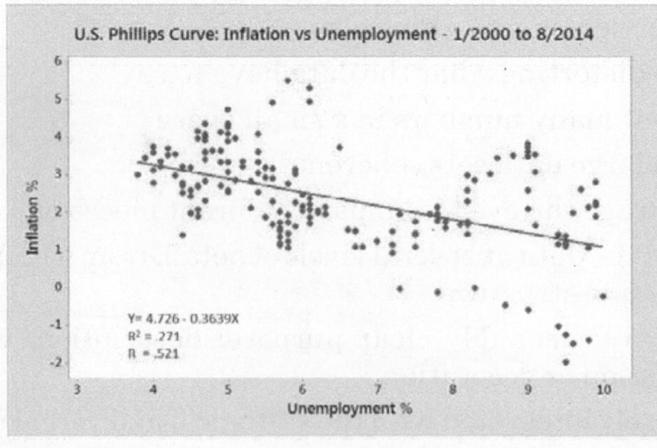

Figure: *A scatterplot illustrating correlation between two variables (inflation and unemployment) measured at points in time.*

Author Stephen Few described eight types of quantitative messages that users may attempt to understand or communicate from a set of data and the associated graphs used to help communicate the message:

1. Time-series: A single variable is captured over a period of time, such as the unemployment rate over a 10-year period. A line chart may be used to demonstrate the trend.
2. Ranking: Categorical subdivisions are ranked in ascending or descending order, such as a ranking of sales performance (the measure) by sales persons (the category, with each sales person a categorical subdivision) during a single period. A bar chart may be used to show the comparison across the sales persons.
3. Part-to-whole: Categorical subdivisions are measured as a ratio to the whole (i.e., a percentage out of 100%). A pie chart or bar chart can show the comparison of ratios, such as the market share represented by competitors in a market.
4. Deviation: Categorical subdivisions are compared again a reference, such as a comparison of actual vs. budget expenses for several departments of a business for a given time period. A bar chart can show comparison of the actual versus the reference amount.
5. Frequency distribution: Shows the number of observations of a particular variable for given interval, such as the number of years in which the stock market return is between intervals such as 0-10%, 11-20%, etc. A histogram, a type of bar chart, may be used for this analysis.
6. Correlation: Comparison between observations represented by two variables (X,Y) to determine if they tend to move in the same or opposite directions. For example, plotting unemployment (X) and inflation (Y) for a sample of months. A scatter plot is typically used for this message.
7. Nominal comparison: Comparing categorical subdivisions in no particular order, such as the sales volume by product code. A bar chart may be used for this comparison.
8. Geographic or geospatial: Comparison of a variable across a map or layout, such as the unemployment rate by state or the number of persons on the various floors of a building. A cartogram is a typical graphic used.

Analysts reviewing a set of data may consider whether some or all of the messages and graphic types above are applicable to their task and audience.

The process of trial and error to identify meaningful relationships and messages in the data is part of exploratory data analysis.

Visual Perception and Data Visualization

A human can distinguish differences in line length, shape orientation, and colour (hue) readily without significant processing effort; these are referred to as "pre-attentive attributes."

For example, it may require significant time and effort ("attentive processing") to identify the number of times the digit "5" appears in a series of numbers; but if that digit is different in size, orientation, or colour, instances of the digit can be noted quickly through pre-attentive processing.

Effective graphics take advantage of pre-attentive processing and attributes and the relative strength of these attributes. For example, since humans can more easily process differences in line length than surface area, it may be more effective to use a bar chart (which takes advantage of line length to show comparison) rather than pie charts (which use surface area to show comparison).

Terminology

Data visualization involves specific terminology, some of which is derived from statistics. For example, author Stephen Few defines two types of data, which are used in combination to support a meaningful analysis or visualization:

- Categorical: Text labels describing the nature of the data, such as "Name" or "Age". This term also covers qualitative (non-numerical) data.
- Quantitative: Numerical measures, such as "25" to represent the age in years.

A table contains quantitative data organised into rows and columns with categorical labels. It is primarily used to lookup specific values. In the example above, the table might have categorical column labels representing the name (a qualitative variable) and age (a quantitative variable), with each row of data representing one person (the sampled experimental unit or category subdivision).

A graph is primarily used to show relationships among data and portrays values encoded as visual objects (e.g., lines, bars, or points). Numerical values are displayed within an area delineated by one or more axes. These axes provide scales (quantitative and categorical) used to label and assign values to the visual objects.

ARIMA Models

In statistics and econometrics, and in particular in time series analysis, an autoregressive integrated moving average (ARIMA) model is a generalization of an autoregressive moving average (ARMA) model. These models are fitted to time series data either to better understand the data or to predict future points in the series (forecasting). They are applied in some cases where data show evidence of non-stationarity, where an initial differencing step (corresponding to the "integrated" part of the model) can be applied to remove the non-stationarity.

The model is generally referred to as an ARIMA(p,d,q) model where parameters p, d, and q are non-negative integers that refer to the order of the autoregressive, integrated, and moving average parts of the model respectively. ARIMA models form an important part of the Box-Jenkins approach to time-series modelling.

When two out of the three terms are zeros, it is a common mistake to refer to the model as ARIMA. The model should be referred to as defined in the parameters, dropping "AR", "I" or "MA" from the acronym describing the model. For example, ARIMA(0,1,0) is I(1), and ARIMA(0,0,1) is MA(1).

Definition

Given a time series of data X_t where t is an integer index and the are real numbers, then an ARMA(p' ,q) model is given by:

$$\left(1-\sum_{i=1}^{p'}\alpha_i L^i\right)X_t=\left(1+\sum_{i=1}^{q}\theta_i L^i\right)\varepsilon_t$$

where L is the lag operator, the α_i are the parameters of the autoregressive part of the model, the are the parameters of the moving average part and the ε_t are error terms. The error terms ε_t are generally assumed to be independent, identically distributed variables sampled from a normal distribution with zero mean.

Assume now that the polynomial $\left(1-\sum_{i=1}^{p'}\alpha_i L^i\right)$ has a unitary root of multiplicity d. Then it can be rewritten as:

$$\left(1-\sum_{i=1}^{p'}\alpha_i L^i\right)=\left(1-\sum_{i=1}^{p'-d}\phi_i L^i\right)(1-L)^d.$$

An ARIMA(p,d,q) process expresses this polynomial factorisation property with p=p'–d, and is given by:

$$\left(1-\sum_{i=1}^{p}\phi_i L^i\right)(1-L)^d X_t = \left(1+\sum_{i=1}^{q}\theta_i L^i\right)\varepsilon_t$$

and thus can be thought as a particular case of an ARMA(p+d,q) process having the autoregressive polynomial with d unit roots. (For this reason, every ARIMA model with d>0 is not wide sense stationary.)

The above can be generalized as follows.

$$\left(1-\sum_{i=1}^{p}\phi_i L^i\right)(1-L)^d X_t = \delta + \left(1+\sum_{i=1}^{q}\theta_i L^i\right)\varepsilon_t$$

This defines an ARIMA(p,d,q) process with drift $\delta/(1-\Sigma\varphi_i)$.

Other Special Forms

The explicit identification of the factorisation of the autoregression polynomial into factors as above, can be extended to other cases, firstly to apply to the moving average polynomial and secondly to include other special factors. For example, having a factor $\left(1-L^s\right)$ in a model is one way of including a non-stationary seasonality of period s into the model; this factor has the effect of re-expressing the data as changes from s periods ago. Another example is the factor $\left(1-\sqrt{3}L+L^2\right)$, which includes a (non-stationary) seasonality of period 2. The effect of the first type of factor is to allow each season's value to drift separately over time, whereas with the second type values for adjacent seasons move together.

Identification and specification of appropriate factors in an ARIMA model can be an important step in modelling as it can allow a reduction in the overall number of parameters to be estimated, while allowing the imposition on the model of types of behaviour that logic and experience suggest should be there.

Forecasts Using ARIMA Models

The ARIMA model can be viewed as a "cascade" of two models. The first is non-stationary:

$$Y_t = (1-L)^d X_t$$

while the second is wide-sense stationary:

$$\left(1-\sum_{i=1}^{p}\phi_i L^i\right)Y_t = \left(1+\sum_{i=1}^{q}\theta_i L^i\right)\varepsilon_t.$$

Now forecasts can be made for the process Y_t, using a generalization of the method of autoregressive forecasting.

Examples

Some well-known special cases arise naturally. For example, an ARIMA(0,1,0) model (or I(1) model) is given by

$$X_t = X_{t-1} + \varepsilon_t$$

—which is simply a random walk.

Variations and Extensions

A number of variations on the ARIMA model are commonly employed. If multiple time series are used then the X_t can be thought of as vectors and a VARIMA model may be appropriate. Sometimes a seasonal effect is suspected in the model; in that case, it is generally better to use a SARIMA (seasonal ARIMA) model than to increase the order of the AR or MA parts of the model. If the time-series is suspected to exhibit long-range dependence, then the d parameter may be allowed to have non-integer values in an autoregressive fractionally integrated moving average model, which is also called a Fractional ARIMA (FARIMA or ARFIMA) model.

Implementations in Statistics Packages

Various packages that apply methodology like Box-Jenkins parameter optimization are available to find the right parameters for the ARIMA model.

- In R, the standard stats package includes an arima function, is documented in "ARIMA Modelling of Time Series". Besides the ARIMA(p,d,q) part, the function also includes seasonal factors, an intercept term, and exogenous variables (xreg, called "external regressors"). The CRAN task view on Time Series is the reference with many more links. The "forecast" package in R can automatically select an ARIMA model for a given time series with the auto.arima() function. The package can also simulate seasonal and non-seasonal ARIMA models with its simulate.Arima() function. It also has a function Arima(), which is a wrapper for the arima from the "stats" package.
- In Python_(programming_language), the "statsmodels" package includes models for time series analysis - univariate time series analysis: AR, ARIMA - vector autoregressive models, VAR and structural VAR - descriptive statistics and process models for time series analysis.

- IBM SPSS includes ARIMA modelling in its Statistics and Modeler statistical packages. The default Expert Modeler feature evaluates a range of seasonal and non-seasonal autoregressive (p), integrated (d), and moving average (q) settings and seven exponential smoothing models. The Expert Modeler can also transform the target time-series data into its square root or natural log. The user also has the option to restrict the Expert Modeler to ARIMA models, or to manually enter ARIMA nonseasonal and seasonal p, d, and q settings without Expert Modeler. Automatic outlier detection is available for seven types of outliers, and the detected outliers will be accommodated in the time-series model if this feature is selected.
- The APO-FCS package in SAP ERP from SAP allows creation and fitting of ARIMA models using the Box-Jenkins methodology.
- SAS includes extensive ARIMA processing in its Econometric and Time Series Analysis system: SAS/ETS.
- Stata includes ARIMA modelling (using its arima command) as of Stata 9.
- SQL Server Analysis Services from Microsoft includes ARIMA as a Data Mining algorithm.
- Mathematica includes ARIMAProcess function.
- EViews has extensive ARIMA and SARIMA capabilities.
- MATLAB's Econometrics Toolbox includes ARIMA models and regression with ARIMA errors
- Julia contains an ARIMA implementation in the TimeModels package

Chapter 3

Spectral Analysis and Filtering

A power spectrum can be calculated from the result of a wavelet transform. Plotting the power spectrum provides a useful graphical representation for analyzing wavelet functions and for defining filters.

A wavelet function can be viewed as a high pass filter, which aproximates a data set (a signal or time series). The result of the wavelet function is the difference between value calculated by the wavelet function and the actual data. The scaling function calculates a smoothed version of the data, which becomes the input for the next iteration of the wavelet function. In the context of filtering, an ideal wavelet/scaling function pair would exactly split the spectrum. For example, if the sampled frequency range is 0 to 1024 Hz, the result of the wavelet function would be signals from 512 to 1025 Hz. The result of the scaling function would be signals from 0 to 511.

There are an infinite variety of wavelet and wavelet scaling functions. The closeness of the approximation provided by the wavelet function depends on the nature of the data. How closely the wavelet and scaling functions approach ideal filters depends on the nature of these functions and how they interact with the data.

Calculating the Wavelet Power Spectrum

Figure given below shows the logical structure of the result of a wavelet transform calculated on a 16-element data set. The element labeled "data average" is the result of the final application of the scaling function. In the case of the Haar wavelet function, this will be the average of the data set. Following the data average are the wavelet coefficient bands, whose size is an increasing power of two (e.g., 2^0, 2^1, 2^2...)

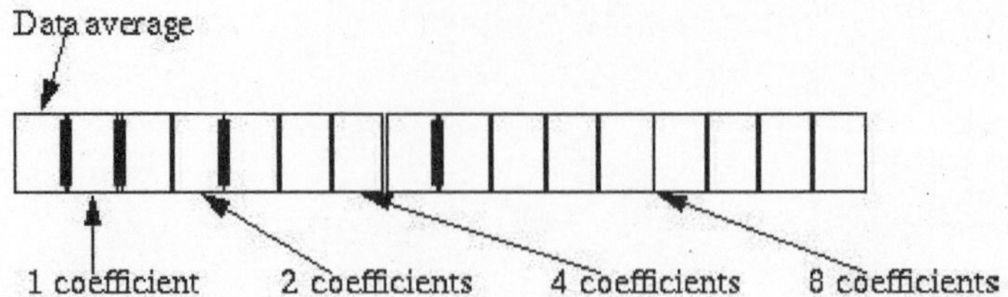

Result of a wavelet transform with coefficients ordered in increasing frequency

If the data set consists of N elements, where N is a power of two, there will be $\log_2(N)$ coefficient bands and one scaling value. For example, in the case of the 16 element wavelet result diagrammed in Figure earlier, there are four bands ($\log_2(16) = 4$). The wavelet power spectrum is calculated by summing the squares of the coefficient values for each band.

$$spectrum[j] = \sum_{k=0}^{2j-1} cj,k^2$$

In the spectrum plots below the average is squared as well. This results in the square of the average at spectrum_0 and the square of the first coefficient band at spectrum.

A stationary data set is a data set where the sample can be viewed as repeating infinitely if the data set were also infinitely extended. An example of a stationary data set is a data set composed of sine and/or cosine waves.

The goal of spectral estimation is to describe the distribution (over frequency) of the power contained in a signal, based on a finite set of data. Estimation of power spectra is useful in a variety of applications, including the detection of signals buried in wideband noise.

Spectral Estimation Method

The various methods of spectrum estimation available in the toolbox are categorized as follows:

- Nonparametric methods
- Parametric methods
- Subspace methods

Nonparametric methods are those in which the PSD is estimated directly from the signal itself. The simplest such method is the periodogram. Other nonparametric techniques such as Welch's method, the multitaper method (MTM) reduce the variance of the periodogram.

Parametric methods are those in which the PSD is estimated from a signal that is assumed to be output of a linear system driven by white noise. Examples are the Yule-Walker autoregressive (AR) method and the Burg method. These methods estimate the PSD by first estimating the parameters (coefficients) of the linear system that hypothetically generates the signal. They tend to produce better results than classical nonparametric methods when the data length of the available signal is relatively short. Parametric methods also produce smoother estimates of the PSD than nonparametric methods, but are subject to error from model misspecification.

Subspace methods, also known as high-resolution methods or super-resolution methods, generate frequency component estimates for a signal based on an eigenanalysis or eigendecomposition of the autocorrelation matrix. Examples are the multiple signal classification (MUSIC) method or the eigenvector (EV) method. These methods are best suited for line spectra — that is, spectra of sinusoidal signals — and are effective in the detection of sinusoids buried in noise, especially when the signal to noise ratios are low. The subspace methods do not yield true PSD estimates: they do not preserve process power between the time and frequency domains, and the autocorrelation sequence cannot be recovered by taking the inverse Fourier transform of the frequency estimate.

All three categories of methods are listed in the table below with the corresponding toolbox function names.

Spectral Estimation Methods/Functions

Method	*Description*	*Functions*
Periodogram	Power spectral density estimate	periodogram
Welch	Averaged periodograms of overlapped, windowed signal sections	pwelch, cpsd, tfestimate, mscohere
Multitaper	Spectral estimate from combination of multiple orthogonal windows (or "tapers")	pmtm
Yule-Walker AR	Autoregressive (AR) spectral estimate of a time-series from its estimated autocorrelation function	pyulear
Burg	Autoregressive (AR) spectral estimation of a time-series by minimization of linear prediction errors	pburg
Covariance	Autoregressive (AR) spectral estimation of a time-series by minimization of the forward prediction errors	pcov
Modified Covariance	Autoregressive (AR) spectral estimation of a time-series by minimization of the forward and backward prediction errors	pmcov
MUSIC	Multiple signal classification	pmusic
Eigenvector	Pseudospectrum estimate	peig

Nonparametric Methods

The following sections discuss the periodogram, modified periodogram, Welch, and multitaper methods of nonparametric estimation, along with the related CPSD function, transfer function estimate, and coherence function.

Periodogram

In general terms, one way of estimating the PSD of a process is to simply find the discrete-time Fourier transform of the samples of the process (usually done on a grid with an FFT) and appropriately scale the magnitude squared of the result. This estimate is called the periodogram.

As an example of the periodogram, consider the following 1001-element signal xn, which consists of two sinusoids plus noise:

```
fs = 1000;                    % Sampling frequency
t = (0:fs)/fs;                % One second worth of samples
A = [1 2];                    % Sinusoid amplitudes (row vector)
f = [150;140];                % Sinusoid frequencies (column vector)
xn = A*sin(2*pi*f*t) + 0.1*randn(size(t));
```

Note: The three last lines illustrate a convenient and general way to express the sum of sinusoids. Together they are equivalent to xn = sin(2*pi*150*t) + 2*sin(2*pi*140*t) + 0.1*randn(size(t));

The periodogram estimate of the PSD can be computed using periodogram. In this case, the data vector is multiplied by a Hamming window to produce a modified periodogram.

```
[Pxx,F] = periodogram(xn,hamming(length(xn)),length(xn),fs);
plot(F,10*log10(Pxx))
xlabel('Hz')
ylabel('dB')
title('Modified Periodogram Power Spectral Density Estimate')
```

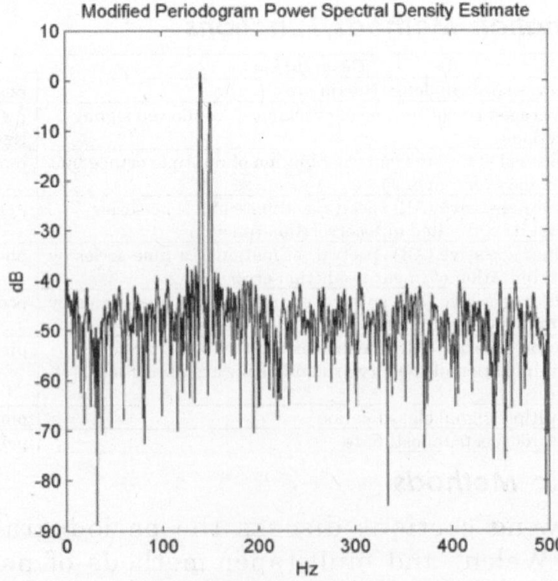

Algorithm. Periodogram computes and scales the output of the FFT to produce the power vs. frequency plot as follows:

1. If the input signal is real-valued, the magnitude of the resulting FFT is symmetric with respect to zero frequency (DC).

   For an even-length FFT, only the first (1 + nfft/2) points are unique. Determine the number of unique values and keep only those unique points.
2. Take the squared magnitudes of the unique FFT values. Scale the squared magnitudes (except for DC) by $2/(F_s N)$, where N is the length of signal prior to any zero padding. Scale the DC value by $1/(F_s N)$.
3. Create a frequency vector from the number of unique points, the nfft and the sampling frequency.
4. Plot the resulting magnitude squared FFT against the frequency.

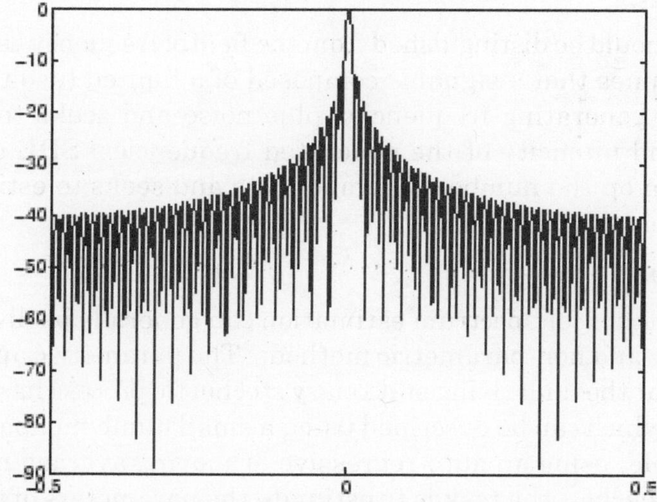

The plot displays a main lobe and several side lobes, the largest of which is approximately 13.5 dB below the mainlobe peak. These lobes account for the effect known as spectral leakage.

While the infinite-length signal has its power concentrated exactly at the discrete frequency points f_k, the windowed (or truncated) signal has a continuum of power "leaked" around the discrete frequency points f_k.

Because the frequency response of a short rectangular window is a much poorer approximation to the Dirac delta function than that of a longer window, spectral leakage is especially evident when data records are short. Consider the following sequence of 100 samples:

```
fs = 1000;                    % Sampling frequency
t = (0:fs/10)/fs;             % One-tenth second worth of samples
A = [1 2];                    % Sinusoid amplitudes
f = [150;140];                % Sinusoid frequencies
xn = A*sin(2*pi*f*t) + 0.1*randn(size(t));
[Pxx,F] = periodogram(xn,rectwin(length(xn)),length(xn),fs);
plot(F,10*log10(Pxx))
```

Spectral Density Estimation

In statistical signal processing, the goal of spectral density estimation (SDE) is to estimate the spectral density (also known as the power spectral density) of a random signal from a sequence of time samples of the signal. Intuitively speaking, the spectral density characterizes the frequency content of the signal. One purpose of estimating the spectral density is to detect any periodicities in the data, by observing peaks at the frequencies corresponding to these periodicities.

SDE should be distinguished from the field of frequency estimation, which assumes that a signal is composed of a limited (usually small) number of generating frequencies plus noise and seeks to find the location and intensity of the generated frequencies. SDE makes no assumption on the number of components and seeks to estimate the whole generating spectrum.

Techniques

Techniques for spectrum estimation can generally be divided into parametric and non-parametric methods. The parametric approaches assume that the underlying stationary stochastic process has a certain structure which can be described using a small number of parameters (for example, using an auto-regressive or moving average model). In these approaches, the task is to estimate the parameters of the model that describes the stochastic process. By contrast, non-parametric approaches explicitly estimate the covariance or the spectrum of the process without assuming that the process has any particular structure.

Following is a partial list of spectral density estimation techniques:

- Periodogram the basic modulus-squared of the Fourier transform
- Bartlett's method is a periodogram spectral estimate formed by averaging the discrete Fourier transform of multiple segments of the signal to reduce variance of the spectral density estimate

- Welch's method a windowed version of Bartlett's method that uses overlapping segments
- Blackman-Tukey is a periodogram smoothing technique to reduce variance of the spectral density estimate
- Autoregressive moving average estimation, based on fitting to an ARMA model to the time series of signal samples; in addition to ARMA, there are separate moving average and autoregressive methods
- Multitaper is a periodogram-based method that uses multiple tapers, or windows, to form independent estimates of the spectral density to reduce variance of the spectral density estimate
- Maximum entropy spectral estimation is an all-poles method useful for SDE when singular spectral features, such as sharp peaks, are expected.
- Least-squares spectral analysis, based on least squares fitting to known frequencies
- Non-uniform discrete Fourier transform is used when the signal samples are unevenly spaced in time
- Singular spectrum analysis is a nonparametric method that uses a singular value decomposition of the covariance matrix to estimate the spectral density
- Short-time Fourier transform

Parametric Estimation

In parametric spectral estimation, one assumes that the signal is modelled by a stationary process which has a spectral density function (SDF) $S(f;a_1,\ldots,a_p)$ that is a function of the frequency *f* and *p* parameters $a_1,\ldots,a_p$. The estimation problem then becomes one of estimating these parameters.

The most common form of parametric SDF estimate uses as a model an autoregressive model $AR(p)$ of order p. A signal sequence $\{Y_t\}$ obeying a zero mean $AR(p)$ process satisfies the equation

$$Y_t = \phi_1 Y_{t-1} + \phi_2 Y_{t-2} + \cdots + \phi_p Y_{t-p} + \epsilon_t,$$

where the $\phi_1,\ldots,\phi_p$ are fixed coefficients and ϵ_t is a white noise process with zero mean and innovation variance σ_p^2. The SDF for this process is

$$S(f;\phi_1,\ldots,\phi_p,\sigma_p^2)=\frac{\sigma_p^2\Delta t}{\left|1-\sum_{k=1}^{p}\phi_k e^{-2i\pi fk\Delta t}\right|^2}\qquad |f|<f_N,$$

with Δt the sampling time interval and f_N the Nyquist frequency.

There are a number of approaches to estimating the parameters $\phi_1,\ldots,\phi_p,\sigma_p^2$ of the process and this the spectral density:

- The Yule-Walker estimators are found by recursively solving the Yule-Walker equations for an $AR(p)$ process
- The Burg estimators are found by treating the Yule-Walker equations as a form of ordinary least squares problem. The Burg estimators are generally considered superior to the Yule-Walker estimators. Burg associated these with maximum entropy spectral estimation.
- The forward-backward least-squares estimators treat the $AR(p)$ process as a regression problem and solves that problem using forward-backward method. They are competitive with the Burg estimators.
- The maximum likelihood estimators assume the white noise is a Gaussian process and estimates the parameters using a maximum likelihood approach. This involves a nonlinear optimization and is more complex than the first three.

Alternative parametric methods include fitting to a moving average model (MA) and to a full autoregressive moving average model (ARMA).

Frequency Estimation

Frequency estimation is the process of estimating the complex frequency components of a signal in the presence of noise given assumptions about the number of the components. This contrasts with the general methods above, which do not make prior assumptions about the components.

Finite Number of Tones

A typical model for a signal $x(n)$ consists of a sum of p complex exponentials in the presence of white noise, $w(n)$

$$x(n)=\sum_{i=1}^{p}A_i e^{jn\omega_i}+w(n).$$

The power spectral density of $x(n)$ is composed of p impulse functions in addition to the spectral density function due to noise.

The most common methods for frequency estimation involve identifying the noise subspace to extract these components. These methods are based on eigen decomposition of the autocorrelation matrix into a signal subspace and a noise subspace. After these subspaces are identified, a frequency estimation function is used to find the component frequencies from the noise subspace. The most popular methods of noise subspace based frequency estimation are Pisarenko's method, the multiple signal classification (MUSIC) method, the eigenvector method, and the minimum norm method.

$$\hat{P}_{PHD}(e^{j\omega}) = \frac{1}{|\mathbf{e}^H \mathbf{v}_{min}|^2}$$

Pisarenko's method

$$\hat{P}_{MU}(e^{j\omega}) = \frac{1}{\sum_{i=p+1}^{M} |\mathbf{e}^H \mathbf{v}_i|^2}$$

MUSIC

$$\hat{P}_{EV}(e^{j\omega}) = \frac{1}{\sum_{i=p+1}^{M} \frac{1}{\lambda_i} |\mathbf{e}^H \mathbf{v}_i|^2},$$

Eigenvector method

$$\hat{P}_{MN}(e^{j\omega}) = \frac{1}{|\mathbf{e}^H \mathbf{a}|^2}; \mathbf{a} = \lambda \mathbf{P}_n \mathbf{u}_1$$

Minimum norm method

Single Tone

If one only wants to estimate the single loudest frequency, one can use a pitch detection algorithm. If the dominant frequency changes over time, then the problem becomes the estimation of the instantaneous frequency as defined in the time–frequency representation. Methods for instantaneous frequency estimation include those based on the Wigner-Ville distribution and higher order ambiguity functions.

If one wants to know all the (possibly complex) frequency components of a received signal (including transmitted signal and noise), one uses a discrete Fourier transform or some other Fourier-related transform.

Time–Frequency Analysis

In signal processing, time–frequency analysis comprises those techniques that study a signal in both the time and frequency domains simultaneously, using various time–frequency representations. Rather than viewing a 1-dimensional signal (a function, real or complex-valued, whose domain is the real line) and some transform (another function whose domain is the real line, obtained from the original via some transform), time–frequency analysis studies a two-dimensional signal – a function whose domain is the two-dimensional real plane, obtained from the signal via a time–frequency transform.

The mathematical motivation for this study is that functions and their transform representation are often tightly connected, and they can be understood better by studying them jointly, as a two-dimensional object, rather than separately. A simple example is that the 4-fold periodicity of the Fourier transform – and the fact that two-fold Fourier transform reverses direction – can be interpreted by considering the Fourier transform as a 90° rotation in the associated time–frequency plane: 4 such rotations yield the identity, and 2 such rotation simply reverse direction (reflection through the origin).

The practical motivation for time–frequency analysis is that classical Fourier analysis assumes that signals are infinite in time or periodic, while many signals in practice are of short duration, and change substantially over their duration. For example, traditional musical instruments do not produce infinite duration sinusoids, but instead begin with an attack, then gradually decay. This is poorly represented by traditional methods, which motivates time–frequency analysis.

One of the most basic forms of time–frequency analysis is the short-time Fourier transform (STFT), but more sophisticated techniques have been developed, notably wavelets.

Need for a Time–Frequency Approach

In signal processing, time–frequency analysis is a body of techniques and methods used for characterizing and manipulating signals whose statistics vary in time, such as transient signals.

It is a generalization and refinement of Fourier analysis, for the case when the signal frequency characteristics are varying with time. Since many signals of interest – such as speech, music, images, and medical signals – have changing frequency characteristics, time–frequency analysis has broad scope of applications.

Whereas the technique of the Fourier transform can be extended to obtain the frequency spectrum of any slowly growing locally integrable signal, this approach requires a complete description of the signal's behaviour over all time. Indeed, one can think of points in the (spectral) frequency domain as smearing together information from across the entire time domain. While mathematically elegant, such a technique is not appropriate for analyzing a signal with indeterminate future behaviour. For instance, one must presuppose some degree of indeterminate future behaviour in any telecommunications systems to achieve non-zero entropy (if one already knows what the other person will say one cannot learn anything).

To harness the power of a frequency representation without the need of a complete characterization in the time domain, one first obtains a time–frequency distribution of the signal, which represents the signal in both the time and frequency domains simultaneously. In such a representation the frequency domain will only reflect the behaviour of a temporally localized version of the signal. This enables one to talk sensibly about signals whose component frequencies vary in time.

For instance rather than using tempered distributions to globally transform the following function into the frequency domain one could instead use these methods to describe it as a signal with a time varying frequency.

$$x(t) = \begin{cases} \cos(\pi t); & t < 10 \\ \cos(3\pi t); & 10 \le t < 20 \\ \cos(2\pi t); & t > 20 \end{cases}$$

Once such a representation has been generated other techniques in time–frequency analysis may then be applied to the signal in order to extract information from the signal, to separate the signal from noise or interfering signals, etc.

Time–Frequency Distribution Functions

Diversity of Time–Frequency Formulations

There are several different ways to formulate a valid time–frequency distribution function, resulting in several well-known time–frequency distributions, such as:

- Short-time Fourier transform (including the Gabor transform),
- Wavelet transform,

- Bilinear time–frequency distribution function (Wigner distribution function),
- Modified Wigner distribution function, Gabor–Wigner distribution function, and so on.

More information about the history and the motivation of development of time–frequency distribution can be found in the entry Time–frequency representation.

Ideal TF Distribution Function

A time–frequency distribution function ideally has the following properties:

1. High clarity to make it easier to be analyzed and interpreted.
2. No cross-term to avoid confusing real components from artifacts or noise.
3. A list of desirable mathematical properties to ensure such methods benefit real-life application.
4. Lower computational complexity to ensure the time needed to represent and process a signal on a time–frequency plane allows real-time implementations.

Below is a brief comparison of some selected time–frequency distribution functions.

	Clarity	*Cross-term*	*Good mathematical properties*	*Computational complexity*
Gabor transform	Worst	No	Worst	Low
Wigner distribution function	Best	Yes	Best	High
Gabor-Wigner distribution function	Good	Almost eliminated	Good	High
Cone-shape distribution function	Good	No(eliminated, in time)	Good	Medium (if recursively defined)

To analyze the signals well, choosing an appropriate time–frequency distribution function is important. Which time–frequency distribution function should be used depends on the application being considered, as shown by reviewing a list of applications. The high clarity of the Wigner distribution function (WDF) obtained for some signals is due to the auto-correlation function inherent in its formulation; however, the latter also causes the cross-term problem.

Therefore, if we want to analyze a single-term signal, using the WDF may be the best approach; if the signal is composed of multiple components, some other methods like the Gabor transform, Gabor-Wigner distribution or Modified B-Distribution functions may be better choices.

To illustrate this, we observe that by Fourier analysis, we can't recognise the two signals $x_1(t)$ and $x_2(t)$ below.

$$x_1(t) = \begin{cases} \cos(\pi t); & t < 10 \\ \cos(3\pi t); & 10 \le t < 20 \\ \cos(2\pi t); & t > 20 \end{cases}$$

$$x_2(t) = \begin{cases} \cos(\pi t); & t < 10 \\ \cos(2\pi t); & 10 \le t < 20 \\ \cos(3\pi t); & t > 20 \end{cases}$$

Thanks to the time–frequency analysis approach, we can still solve this problem of correctly identifying the two different signals.

Signal Processing Applications

The following applications need not only the time–frequency distribution functions but also some operations to the signal. The Linear canonical transform (LCT) is really helpful. By LCTs, the shape and location on the time–frequency plane of a signal can be in the arbitrary form that we want it to be. For example, the LCTs can shift the time–frequency distribution to any location, dilate it in the horizontal and vertical direction without changing its area on the plane, shear (or twist) it, and rotate it (Fractional Fourier transform). This powerful operation, LCT, make it more flexible to analyze and apply the time–frequency distributions. Here we list some applications of time–frequency analysis.

Instantaneous Frequency Estimation

The definition of instantaneous frequency is the time rate of change of phase, or

$$\frac{1}{2\pi}\frac{d}{dt}\phi(t),$$

where $\phi(t)$ is the instantaneous phase of a signal. We can know the instantaneous frequency from the time–frequency plane directly if the image is clear enough. Because the high clarity is critical, we often use WDF to analyze it.

TF Filtering and Signal Decomposition

The goal of filter design is to remove the undesired component of a signal. Conventionally, we can just filter in the time domain or in the frequency domain individually as shown below.

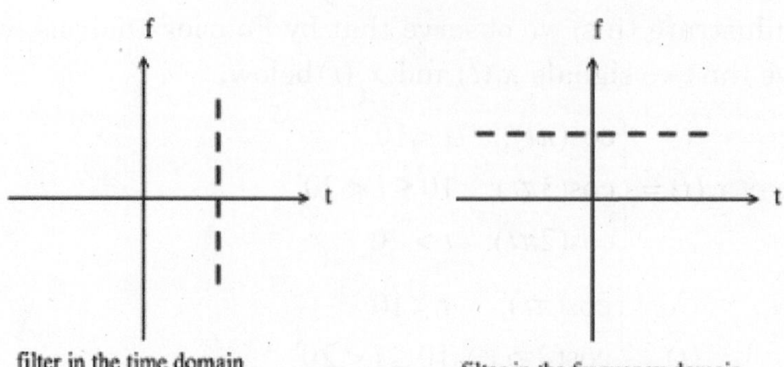

The filtering methods mentioned above can't work well for every signal which may overlap in the time domain or in the frequency domain. By using the time–frequency distribution function, we can filter in the Euclidian time–frequency domain or in the fractional domain by employing the fractional Fourier transform. An example is shown below.

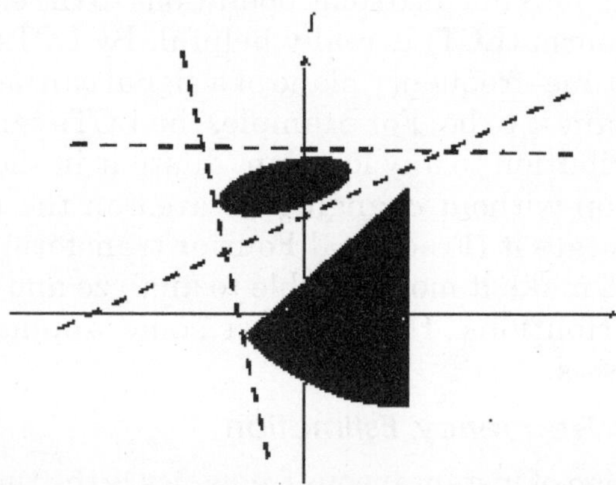

Filter design in time–frequency analysis always deals with signals composed of multiple components, so one cannot use WDF due to cross-term. The Gabor transform, Gabor-Wigner distribution function, or Cohen's class distribution function may be better choices.

The concept of signal decomposition relates to the need to separate one component from the others in a signal; this can be achieved through a filtering operation which require a filter design stage. Such filtering is traditionally done in the time domain or in the frequency domain; however, this may not be possible in the case of non-stationary signals that are multicomponent as such components could overlap in both the time domain and also in the frequency domain; as a consequence, the only possible way to achieve component separation and therefore a signal decomposition is to implement a time–frequency filter.

Sampling Theory

By the Nyquist–Shannon sampling theorem, we can conclude that the minimum number of sampling points without aliasing is equivalent to the area of the time–frequency distribution of a signal. (This is actually just an approximation, because the TF area of any signal is infinite.) Below is an example before and after we combine the sampling theory with the time–frequency distribution:

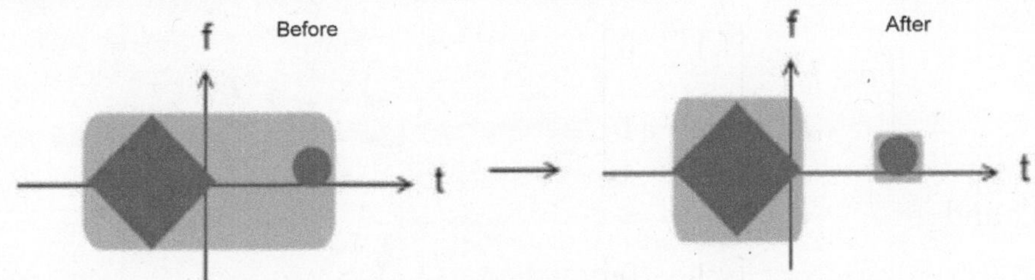

It is noticeable that the number of sampling points decreases after we apply the time–frequency distribution.

When we use the WDF, there might be the cross-term problem (also called interference). On the other hand, using Gabor transform causes an improvement in the clarity and readability of the representation, therefore improving its interpretation and application to practical problems.

Consequently, when the signal we tend to sample is composed of single component, we use the WDF; however, if the signal consists of more than one component, using the Gabor transform, Gabor-Wigner distribution function, or other reduced interference TFDs may achieve better results.

The Balian–Low theorem formalizes this, and provides a bound on the minimum number of time–frequency samples needed.

Electromagnetic Wave Propagation

We can represent an electromagnetic wave in the form of a 2 by 1 matrix

$$\begin{bmatrix} x \\ y \end{bmatrix},$$

which is similar to the time–frequency plane. When electromagnetic wave propagates through free-space, the Fresnel diffraction occurs. We can operate with the 2 by 1 matrix

$$\begin{bmatrix} x \\ y \end{bmatrix}$$

by LCT with parameter matrix

$$\begin{bmatrix} a & b \\ c & d \end{bmatrix} = \begin{bmatrix} 1 & \lambda z \\ 0 & 1 \end{bmatrix},$$

where z is the propagation distance and λ is the wavelength. When electromagnetic wave pass through a spherical lens or be reflected by a disk, the parameter matrix should be

$$\begin{bmatrix} a & b \\ c & d \end{bmatrix} = \begin{bmatrix} 1 & 0 \\ \dfrac{-1}{\lambda f} & 1 \end{bmatrix}$$

and

$$\begin{bmatrix} a & b \\ c & d \end{bmatrix} = \begin{bmatrix} 1 & 0 \\ \dfrac{1}{\lambda R} & 1 \end{bmatrix}$$

respectively, where f is the focal length of the lens and R is the radius of the disk. These corresponding results can be obtained from

$$\begin{bmatrix} a & b \\ c & d \end{bmatrix}\begin{bmatrix} x \\ y \end{bmatrix}.$$

Optics, Acoustics, and Biomedicine

Light is a kind of electromagnetic wave, so we apply the time–frequency analysis to optics in the same way as to electromagnetic wave propagation. In the same way, a characteristic of acoustic signals is that, often, its frequency varies really severely with time. Because the acoustic signals usually contain a lot of data, it is suitable to use simpler TFDs such as the Gabor transform to analyze the acoustic signals due to the lower computational complexity. If speed is not an issue, then a detailed comparison with well defined criteria should be made before selecting a particular TFD. Another approach is to define a signal dependent TFD that is adapted to the data. In biomedicine, one can use time–frequency distribution to analyze the electromyography (EMG), Electroencephalography (EEG), Electrocardiogram (ECG) or otoacoustic emissions (OAEs).

Filtering

In signal processing, a filter is a device or process that removes from a signal some unwanted component or feature. Filtering is a class of signal processing, the defining feature of filters being the complete or partial suppression of some aspect of the signal. Most often, this means removing some frequencies and not others in order to suppress interfering signals and reduce background noise. However, filters do not exclusively act in the frequency domain; especially in the field of image processing many other targets for filtering exist. Correlations can be removed for certain frequency components and not for others without having to act in the frequency domain.

There are many different bases of classifying filters and these overlap in many different ways; there is no simple hierarchical classification. Filters may be:

- linear or non-linear
- time-invariant or time-variant, also known as shift invariance. If the filter operates in a spatial domain then the characterization is space invariance.
- causal or not-causal: depending if present output depends or not on "future" input; of course, for time related signals processed in real-time all the filters are causal; it is not necessarily so for filters acting on space-related signals or for deferred-time processing of time-related signals.
- analog or digital
- discrete-time (sampled) or continuous-time
- passive or active type of continuous-time filter
- infinite impulse response (IIR) or finite impulse response (FIR) type of discrete-time or digital filter.

Linear Continuous-Time Filters

Linear continuous-time circuit is perhaps the most common meaning for filter in the signal processing world, and simply "filter" is often taken to be synonymous. These circuits are generally designed to remove certain frequencies and allow others to pass. Circuits that perform this function are generally linear in their response, or at least approximately so. Any nonlinearity would potentially result in the output signal containing frequency components not present in the input signal.

The modern design methodology for linear continuous-time filters is called network synthesis. Some important filter families designed in this way are:

- Chebyshev filter, has the best approximation to the ideal response of any filter for a specified order and ripple.

- Butterworth filter, has a maximally flat frequency response.
- Bessel filter, has a maximally flat phase delay.
- Elliptic filter, has the steepest cutoff of any filter for a specified order and ripple.

The difference between these filter families is that they all use a different polynomial function to approximate to the ideal filter response. This results in each having a different transfer function.

Another older, less-used methodology is the image parameter method. Filters designed by this methodology are archaically called "wave filters". Some important filters designed by this method are:

- Constant k filter, the original and simplest form of wave filter.
- m-derived filter, a modification of the constant k with improved cutoff steepness and impedance matching.

Terminology

Some terms used to describe and classify linear filters:

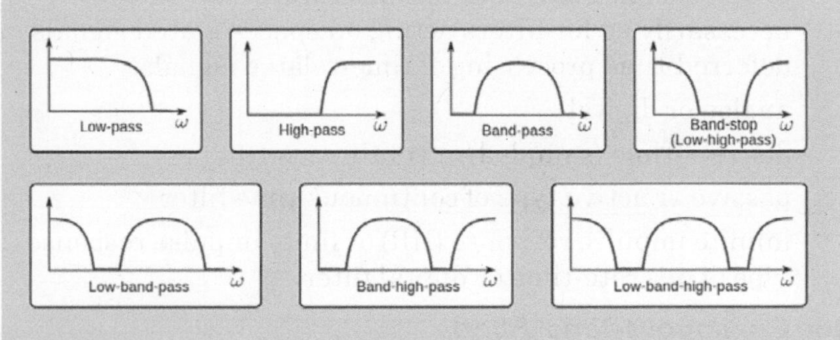

- The frequency response can be classified into a number of different bandforms describing which frequency bands the filter passes (the passband) and which it rejects (the stopband):
    - o Low-pass filter – low frequencies are passed, high frequencies are attenuated.
    - o High-pass filter – high frequencies are passed, low frequencies are attenuated.
    - o Band-pass filter – only frequencies in a frequency band are passed.
    - o Band-stop filter or band-reject filter – only frequencies in a frequency band are attenuated.
    - o Notch filter – rejects just one specific frequency - an extreme band-stop filter.

 - o Comb filter – has multiple regularly spaced narrow passbands giving the bandform the appearance of a comb.
 - o All-pass filter – all frequencies are passed, but the phase of the output is modified.
- Cutoff frequency is the frequency beyond which the filter will not pass signals. It is usually measured at a specific attenuation such as 3dB.
- Roll-off is the rate at which attenuation increases beyond the cut-off frequency.
- Transition band, the (usually narrow) band of frequencies between a passband and stopband.
- Ripple is the variation of the filter's insertion loss in the passband.
- The order of a filter is the degree of the approximating polynomial and in passive filters corresponds to the number of elements required to build it. Increasing order increases roll-off and brings the filter closer to the ideal response.

Technologies

Filters can be built in a number of different technologies. The same transfer function can be realised in several different ways, that is the mathematical properties of the filter are the same but the physical properties are quite different. Often the components in different technologies are directly analogous to each other and fulfill the same role in their respective filters. For instance, the resistors, inductors and capacitors of electronics correspond respectively to dampers, masses and springs in mechanics. Likewise, there are corresponding components in distributed element filters.

- Electronic filters were originally entirely passive consisting of resistance, inductance and capacitance. Active technology makes design easier and opens up new possibilities in filter specifications.
- Digital filters operate on signals represented in digital form. The essence of a digital filter is that it directly implements a mathematical algorithm, corresponding to the desired filter transfer function, in its programming or microcode.
- Mechanical filters are built out of mechanical components. In the vast majority of cases they are used to process an electronic signal and transducers are provided to convert this to and from a mechanical vibration. However, examples do exist of filters

that have been designed for operation entirely in the mechanical domain.

- Distributed element filters are constructed out of components made from small pieces of transmission line or other distributed elements. There are structures in distributed element filters that directly correspond to the lumped elements of electronic filters, and others that are unique to this class of technology.
- Waveguide filters consist of waveguide components or components inserted in the waveguide. Waveguides are a class of transmission line and many structures of distributed element filters, for instance the stub (electronics), can be implemented in waveguides also.
- Crystal filters use quartz crystals as resonators, or some other piezoelectric material.
- Acoustic filters
- Optical filters were originally developed for purposes other than signal processing such as lighting and photography. With the rise of optical fibre technology, however, optical filters increasingly find signal processing applications and signal processing filter terminology, such as longpass and shortpass, are entering the field.

The Transfer Function

The transfer function of a filter is most often defined in the domain of the complex frequencies. The back and forth passage to/from this domain is operated by the Laplace transform and its inverse (therefore, here below, the term "input signal" shall be understood as "the Laplace transform of" (the time representation of) the input signal, and so on).

The transfer function $H(s)$ of a filter is the ratio of the output signal $Y(s)$ to that of the input signal $X(s)$ as a function of the complex frequency S:

$$H(s) = \frac{Y(s)}{X(s)}$$

with $s = \sigma + j\omega$.

The transfer function of all linear time-invariant filters generally share certain characteristics:

- For filters which are constructed of discrete components, their transfer function must be the ratio of two polynomials in , i.e. a rational function of . The order of the transfer function will be the highest power of encountered in either the numerator or the denominator.

- The polynomials of the transfer function will all have real coefficients. Therefore, the poles and zeroes of the transfer function will either be real or occur in complex conjugate pairs.
- Since the filters are assumed to be stable, the real part of all poles (i.e. zeroes of the denominator) will be negative, i.e. they will lie in the left half-plane in complex frequency space.

Distributed element filters do not, in general, produce rational functions but can often approximate to them.

The proper construction of a transfer function involves the Laplace transform, and therefore it is needed to assume null initial conditions, because

$$\mathcal{L}\left\{\frac{df}{dt}\right\} = s\cdot\mathcal{L}\{f(t)\} - f(0),$$

And when f(0)=0 we can get rid of the constants and use the usual expression

$$\mathcal{L}\left\{\frac{df}{dt}\right\} = s\cdot\mathcal{L}\{f(t)\}$$

An alternative to transfer functions is to give the behaviour of the filter as a convolution. The convolution theorem, which holds for Laplace transforms, guarantees equivalence with transfer functions.

Classification

Filters may be specified by family and bandform. A filter's family is specified by the approximating polynomial used and each leads to certain characteristics of the transfer function of the filter. Some common filter families and their particular characteristics are:

- Butterworth filter – no gain ripple in pass band and stop band, slow cutoff
- Chebyshev filter (Type I) – no gain ripple in stop band, moderate cutoff
- Chebyshev filter (Type II) – no gain ripple in pass band, moderate cutoff
- Bessel filter – no group delay ripple, no gain ripple in both bands, slow gain cutoff
- Elliptic filter – gain ripple in pass and stop band, fast cutoff
- Optimum "L" filter

- Gaussian filter – no ripple in response to step function
- Hourglass filter
- Raised-cosine filter

Each family of filters can be specified to a particular order. The higher the order, the more the filter will approach the "ideal" filter; but also the longer the impulse response is and the longer the latency will be. An ideal filter has full transmission in the pass band, complete attenuation in the stop band, and an abrupt transition between the two bands, but this filter has infinite order (i.e., the response cannot be expressed as a linear differential equation with a finite sum) and infinite latency (i.e., its compact support in the Fourier transform forces its time response to be ever lasting).

Here is an image comparing Butterworth, Chebyshev, and elliptic filters. The filters in this illustration are all fifth-order low-pass filters. The particular implementation – analog or digital, passive or active – makes no difference; their output would be the same.

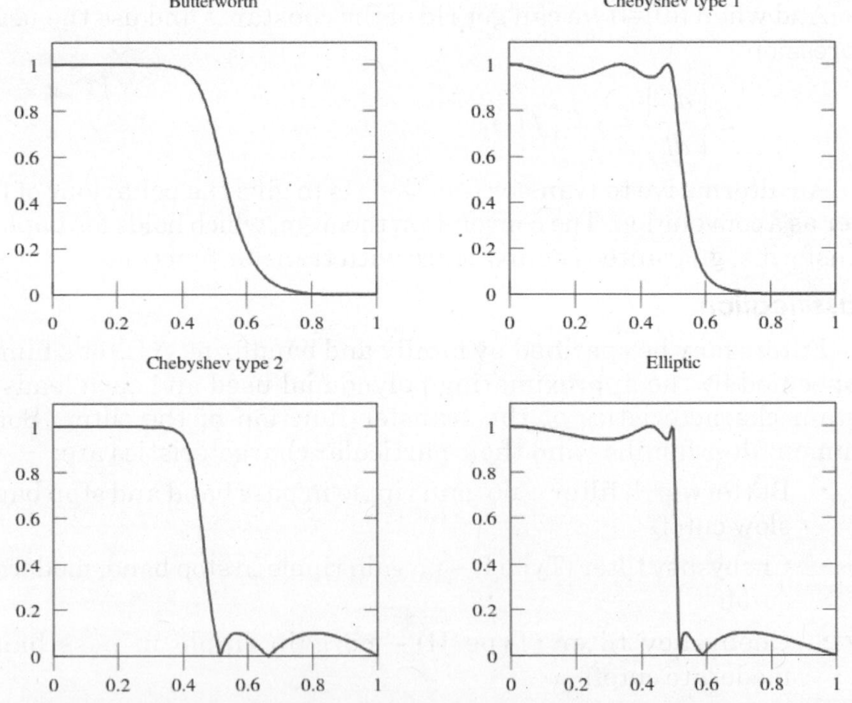

As is clear from the image, elliptic filters are sharper than all the others, but they show ripples on the whole bandwidth.

Any family can be used to implement a particular bandform of which frequencies are transmitted, and which, outside the passband, are more or less attenuated. The transfer function completely specifies the behaviour of a linear filter, but not the particular technology used to implement it. In other words, there are a number of different ways of achieving a particular transfer function when designing a circuit. A particular bandform of filter can be obtained by transformation of a prototype filter of that family.

Impedance Matching

Impedance matching structures invariably take on the form of a filter, that is, a network of non-dissipative elements. For instance, in a passive electronics implementation, it would likely take the form of a ladder topology of inductors and capacitors. The design of matching networks shares much in common with filters and the design invariably will have a filtering action as an incidental consequence. Although the prime purpose of a matching network is not to filter, it is often the case that both functions are combined in the same circuit. The need for impedance matching does not arise while signals are in the digital domain.

Similar comments can be made regarding power dividers and directional couplers. When implemented in a distributed element format, these devices can take the form of a distributed element filter. There are four ports to be matched and widening the bandwidth requires filter-like structures to achieve this. The inverse is also true: distributed element filters can take the form of coupled lines.

Kalman Filter

Kalman filtering, also known as linear quadratic estimation (LQE), is an algorithm that uses a series of measurements observed over time, containing noise (random variations) and other inaccuracies, and produces estimates of unknown variables that tend to be more precise than those based on a single measurement alone. More formally, the Kalman filter operates recursively on streams of noisy input data to produce a statistically optimal estimate of the underlying system state. The filter is named for Rudolf (Rudy) E. Kálmán, one of the primary developers of its theory.

The Kalman filter has numerous applications in technology. A common application is for guidance, navigation and control of vehicles, particularly aircraft and spacecraft. Furthermore, the Kalman filter is a widely applied concept in time series analysis used in fields such as signal processing and econometrics.

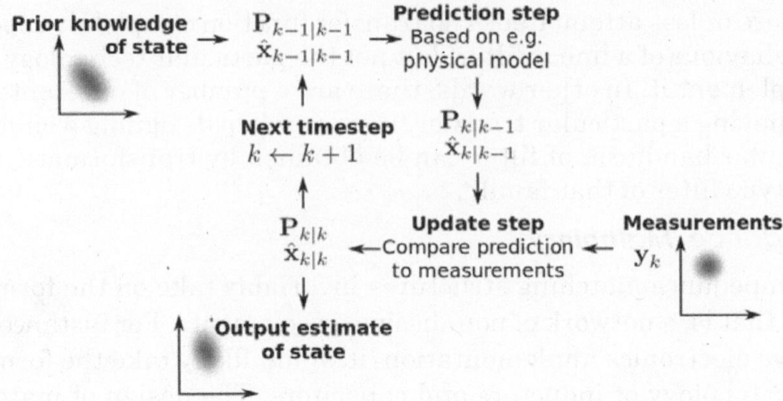

Figure: *The Kalman filter keeps track of the estimated state of the system and the variance or uncertainty of the estimate. The estimate is updated using a state transition model and measurements.* $\hat{x}_{k|k-1}$ *denotes the estimate of the system's state at time step k before the k-th measurement* y_k *has been taken into account;* $P_{k|k-1}$ *is the corresponding uncertainty.*

The algorithm works in a two-step process. In the prediction step, the Kalman filter produces estimates of the current state variables, along with their uncertainties.

Once the outcome of the next measurement (necessarily corrupted with some amount of error, including random noise) is observed, these estimates are updated using a weighted average, with more weight being given to estimates with higher certainty. Because of the algorithm's recursive nature, it can run in real time using only the present input measurements and the previously calculated state and its uncertainty matrix; no additional past information is required.

It is a common misconception that the Kalman filter assumes that all error terms and measurements are Gaussian distributed. Kalman's original paper derived the filter using orthogonal projection theory to show that the covariance is minimized, and this result does not require any assumption, e.g., that the errors are Gaussian. He then showed that the filter yields the exact conditional probability estimate in the special case that all errors are Gaussian-distributed.

Extensions and generalizations to the method have also been developed, such as the extended Kalman filter and the unscented Kalman filter which work on nonlinear systems. The underlying model is a Bayesian model similar to a hidden Markov model but where the state space of the latent variables is continuous and where all latent and observed variables have Gaussian distributions.

Naming and Historical Development

The filter is named after Hungarian émigré Rudolf E. Kálmán, although Thorvald Nicolai Thiele and Peter Swerling developed a similar algorithm earlier. Richard S. Bucy of the University of Southern California contributed to the theory, leading to it often being called the Kalman–Bucy filter. Stanley F. Schmidt is generally credited with developing the first implementation of a Kalman filter. It was during a visit by Kalman to the NASA Ames Research Centre that he saw the applicability of his ideas to the problem of trajectory estimation for the Apollo program, leading to its incorporation in the Apollo navigation computer. This Kalman filter was first described and partially developed in technical papers by Swerling (1958), Kalman (1960) and Kalman and Bucy (1961).

Kalman filters have been vital in the implementation of the navigation systems of U.S. Navy nuclear ballistic missile submarines, and in the guidance and navigation systems of cruise missiles such as the U.S. Navy's Tomahawk missile and the U.S. Air Force's Air Launched Cruise Missile. It is also used in the guidance and navigation systems of the NASA Space Shuttle and the attitude control and navigation systems of the International Space Station.

This digital filter is sometimes called the Stratonovich–Kalman–Bucy filter because it is a special case of a more general, non-linear filter developed somewhat earlier by the Soviet mathematician Ruslan L. Stratonovich. In fact, some of the special case linear filter's equations appeared in these papers by Stratonovich that were published before summer 1960, when Kalman met with Stratonovich during a conference in Moscow.

Overview of the Calculation

The Kalman filter uses a system's dynamics model (e.g., physical laws of motion), known control inputs to that system, and multiple sequential measurements (such as from sensors) to form an estimate of the system's varying quantities (its state) that is better than the estimate obtained by using any one measurement alone. As such, it is a common sensor fusion and data fusion algorithm.

All measurements and calculations based on models are estimates to some degree. Noisy sensor data, approximations in the equations that describe how a system changes, and external factors that are not accounted for introduce some uncertainty about the inferred values for a system's state. The Kalman filter averages a prediction of a system's state with a new measurement using a weighted average.

The purpose of the weights is that values with better (i.e., smaller) estimated uncertainty are "trusted" more. The weights are calculated from the covariance, a measure of the estimated uncertainty of the prediction of the system's state.

The result of the weighted average is a new state estimate that lies between the predicted and measured state, and has a better estimated uncertainty than either alone. This process is repeated every time step, with the new estimate and its covariance informing the prediction used in the following iteration. This means that the Kalman filter works recursively and requires only the last "best guess", rather than the entire history, of a system's state to calculate a new state.

Because the certainty of the measurements is often difficult to measure precisely, it is common to discuss the filter's behaviour in terms of gain. The Kalman gain is a function of the relative certainty of the measurements and current state estimate, and can be "tuned" to achieve particular performance. With a high gain, the filter places more weight on the measurements, and thus follows them more closely. With a low gain, the filter follows the model predictions more closely, smoothing out noise but decreasing the responsiveness. At the extremes, a gain of one causes the filter to ignore the state estimate entirely, while a gain of zero causes the measurements to be ignored.

When performing the actual calculations for the filter (as discussed below), the state estimate and covariances are coded into matrices to handle the multiple dimensions involved in a single set of calculations. This allows for representation of linear relationships between different state variables (such as position, velocity, and acceleration) in any of the transition models or covariances.

Example Application

As an example application, consider the problem of determining the precise location of a truck. The truck can be equipped with a GPS unit that provides an estimate of the position within a few meters. The GPS estimate is likely to be noisy; readings 'jump around' rapidly, though always remaining within a few meters of the real position.

In addition, since the truck is expected to follow the laws of physics, its position can also be estimated by integrating its velocity over time, determined by keeping track of wheel revolutions and the angle of the steering wheel. This is a technique known as dead reckoning. Typically, dead reckoning will provide a very smooth estimate of the truck's position, but it will drift over time as small errors accumulate.

In this example, the Kalman filter can be thought of as operating in two distinct phases: predict and update. In the prediction phase, the truck's old position will be modified according to the physical laws of motion (the dynamic or "state transition" model) plus any changes produced by the accelerator pedal and steering wheel. Not only will a new position estimate be calculated, but a new covariance will be calculated as well. Perhaps the covariance is proportional to the speed of the truck because we are more uncertain about the accuracy of the dead reckoning estimate at high speeds but very certain about the position when moving slowly. Next, in the update phase, a measurement of the truck's position is taken from the GPS unit.

Along with this measurement comes some amount of uncertainty, and its covariance relative to that of the prediction from the previous phase determines how much the new measurement will affect the updated prediction. Ideally, if the dead reckoning estimates tend to drift away from the real position, the GPS measurement should pull the position estimate back towards the real position but not disturb it to the point of becoming rapidly changing and noisy.

Technical Description and Context

The Kalman filter is an efficient recursive filter that estimates the internal state of a linear dynamic system from a series of noisy measurements. It is used in a wide range of engineering and econometric applications from radar and computer vision to estimation of structural macroeconomic models, and is an important topic in control theory and control systems engineering. Together with the linear-quadratic regulator (LQR), the Kalman filter solves the linear-quadratic-Gaussian control problem (LQG). The Kalman filter, the linear-quadratic regulator and the linear-quadratic-Gaussian controller are solutions to what arguably are the most fundamental problems in control theory.

In most applications, the internal state is much larger (more degrees of freedom) than the few "observable" parameters which are measured. However, by combining a series of measurements, the Kalman filter can estimate the entire internal state.

In Dempster–Shafer theory, each state equation or observation is considered a special case of a linear belief function and the Kalman filter is a special case of combining linear belief functions on a join-tree or Markov tree. Additional approaches include belief filters which use Bayes or evidential updates to the state equations.

A wide variety of Kalman filters have now been developed, from Kalman's original formulation, now called the "simple" Kalman filter, the Kalman–Bucy filter, Schmidt's "extended" filter, the information filter, and a variety of "square-root" filters that were developed by Bierman, Thornton and many others. Perhaps the most commonly used type of very simple Kalman filter is the phase-locked loop, which is now ubiquitous in radios, especially frequency modulation (FM) radios, television sets, satellite communications receivers, outer space communications systems, and nearly any other electronic communications equipment.

Underlying Dynamic System Model

The Kalman filters are based on linear dynamic systems discretized in the time domain. They are modelled on a Markov chain built on linear operators perturbed by errors that may include Gaussian noise. The state of the system is represented as a vector of real numbers. At each discrete time increment, a linear operator is applied to the state to generate the new state, with some noise mixed in, and optionally some information from the controls on the system if they are known. Then, another linear operator mixed with more noise generates the observed outputs from the true ("hidden") state. The Kalman filter may be regarded as analogous to the hidden Markov model, with the key difference that the hidden state variables take values in a continuous space (as opposed to a discrete state space as in the hidden Markov model). There is a strong duality between the equations of the Kalman Filter and those of the hidden Markov model. A review of this and other models is given in Roweis and Ghahramani (1999) and Hamilton (1994).

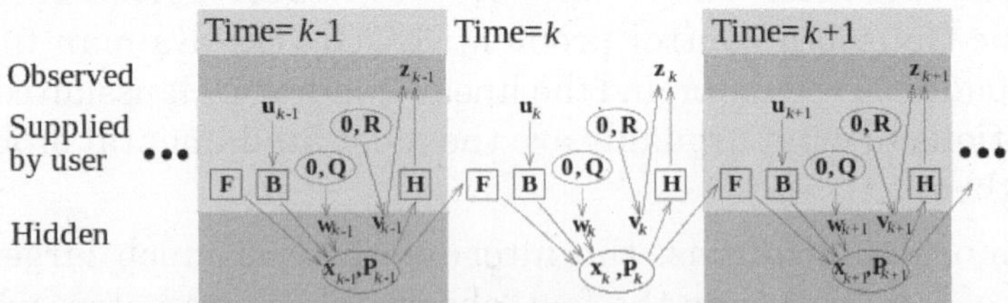

***Figure:** Model underlying the Kalman filter. Squares represent matrices. Ellipses represent multivariate normal distributions (with the mean and covariance matrix enclosed). Unenclosed values are vectors. In the simple case, the various matrices are constant with time, and thus the subscripts are dropped, but the Kalman filter allows any of them to change each time step.*

In order to use the Kalman filter to estimate the internal state of a process given only a sequence of noisy observations, one must model the process in accordance with the framework of the Kalman filter.

This means specifying the following matrices: F_k, the state-transition model; H_k, the observation model; Q_k, the covariance of the process noise; R_k, the covariance of the observation noise; and sometimes B_k, the control-input model, for each time-step, k, as described above.

The Kalman filter model assumes the true state at time k is evolved from the state at (k – 1) according to

$$\mathbf{x}_k = \mathbf{F}_k \mathbf{x}_{k-1} + \mathbf{B}_k \mathbf{u}_k + \mathbf{w}_k$$

where

- F_k is the state transition model which is applied to the previous state x_{k-1};
- B_k is the control-input model which is applied to the control vector u_k;
- w_k is the process noise which is assumed to be drawn from a zero mean multivariate normal distribution with covariance Q_k.

$$\mathbf{w}_k \sim N(0, \mathbf{Q}_k)$$

At time k an observation (or measurement) z_k of the true state x_k is made according to

$$\mathbf{z}_k = \mathbf{H}_k \mathbf{x}_k + \mathbf{v}_k$$

where H_k is the observation model which maps the true state space into the observed space and v_k is the observation noise which is assumed to be zero mean Gaussian white noise with covariance R_k.

$$\mathbf{v}_k \sim N(0, \mathbf{R}_k)$$

The initial state, and the noise vectors at each step $\{x_0, w_1, ..., w_k, v_1 ... v_k\}$ are all assumed to be mutually independent.

Many real dynamical systems do not exactly fit this model. In fact, unmodelled dynamics can seriously degrade the filter performance, even when it was supposed to work with unknown stochastic signals as inputs.

The reason for this is that the effect of unmodelled dynamics depends on the input, and, therefore, can bring the estimation algorithm to instability (it diverges).

On the other hand, independent white noise signals will not make the algorithm diverge. The problem of separating between measurement noise and unmodelled dynamics is a difficult one and is treated in control theory under the framework of robust control.

Details

The Kalman filter is a recursive estimator. This means that only the estimated state from the previous time step and the current measurement are needed to compute the estimate for the current state. In contrast to batch estimation techniques, no history of observations and/or estimates is required. In what follows, the notation $\hat{\mathbf{x}}_{n|m}$ represents the estimate of x at time n given observations up to, and including at time $m \leq n$.

The state of the filter is represented by two variables:

- $\hat{\mathbf{x}}_{k|k}$, the a posteriori state estimate at time k given observations up to and including at time k;
- $\mathbf{P}_{k|k}$, the a posteriori error covariance matrix (a measure of the estimated accuracy of the state estimate).

The Kalman filter can be written as a single equation, however it is most often conceptualized as two distinct phases: "Predict" and "Update". The predict phase uses the state estimate from the previous timestep to produce an estimate of the state at the current timestep. This predicted state estimate is also known as the a priori state estimate because, although it is an estimate of the state at the current timestep, it does not include observation information from the current timestep. In the update phase, the current a priori prediction is combined with current observation information to refine the state estimate. This improved estimate is termed the a posteriori state estimate.

Typically, the two phases alternate, with the prediction advancing the state until the next scheduled observation, and the update incorporating the observation. However, this is not necessary; if an observation is unavailable for some reason, the update may be skipped and multiple prediction steps performed. Likewise, if multiple independent observations are available at the same time, multiple update steps may be performed (typically with different observation matrices H_k).

Predict

Predicted (a priori) state estimate

Predicted (a priori) estimate covariance

Update

Innovation or measurement residual

Innovation (or residual) covariance

Optimal Kalman gain

Updated (a posteriori) state estimate

Updated (a posteriori) estimate covariance

The formula for the updated estimate and covariance above is only valid for the optimal Kalman gain. Usage of other gain values require a more complex formula found in the derivations section.

Invariants

If the model is accurate, and the values for $\hat{\mathbf{x}}_{0|0}$ and $\mathbf{P}_{0|0}$ accurately reflect the distribution of the initial state values, then the following invariants are preserved: (all estimates have a mean error of zero)

- $\mathrm{E}[\mathbf{x}_k - \hat{\mathbf{x}}_{k|k}] = \mathrm{E}[\mathbf{x}_k - \hat{\mathbf{x}}_{k|k-1}] = 0$
- $\mathrm{E}[\tilde{\mathbf{y}}_k] = 0$

where $\mathrm{E}[\xi]$ is the expected value of ξ, and covariance matrices accurately reflect the covariance of estimates

- $\mathbf{P}_{k|k} = \mathrm{cov}(\mathbf{x}_k - \hat{\mathbf{x}}_{k|k})$
- $\mathbf{P}_{k|k-1} = \mathrm{cov}(\mathbf{x}_k - \hat{\mathbf{x}}_{k|k-1})$
- $\mathbf{S}_k = \mathrm{cov}(\tilde{\mathbf{y}}_k)$

Estimation of the Noise Covariances Q_k and R_k

Practical implementation of the Kalman Filter is often difficult due to the inability in getting a good estimate of the noise covariance matrices Q_k and R_k. Extensive research has been done in this field to estimate these covariances from data. One of the more promising and practical approaches to do this is the Autocovariance Least-Squares (ALS) technique that uses the time-lagged autocovariances of routine operating data to estimate the covariances. The GNU Octave and Matlab code used to calculate the noise covariance matrices using the ALS technique is available online under the GNU General Public License license.

Optimality and Performance

It is known from the theory that the Kalman filter is optimal in case that a) the model perfectly matches the real system, b) the entering noise is white and c) the covariances of the noise are exactly known. Several methods for the noise covariance estimation have been proposed during past decades. One, ALS, was mentioned in the previous paragraph. After the covariances are identified, it is useful to evaluate the performance of the filter, i.e. whether it is possible to improve the state estimation quality. It is well known that, if the Kalman filter works optimally, the innovation sequence (the output prediction error) is a white noise. The whiteness property reflects the state estimation quality. For evaluation of the filter performance it is necessary to inspect the whiteness property of the innovations. Several different methods can be used for this purpose. Three optimality tests with numerical examples are described in

Example Application, Technical

Consider a truck on perfectly frictionless, infinitely long straight rails. Initially the truck is stationary at position 0, but it is buffeted this way and that by random acceleration. We measure the position of the truck every Δt seconds, but these measurements are imprecise; we want to maintain a model of where the truck is and what its velocity is. We show here how we derive the model from which we create our Kalman filter.

Since F, H, R and Q are constant, their time indices are dropped.

The position and velocity of the truck are described by the linear state space

$$\mathbf{x}_k = \begin{bmatrix} x \\ \dot{x} \end{bmatrix}$$

where x is the velocity, that is, the derivative of position with respect to time. We assume that between the (k – 1) and k timestep the truck undergoes a constant acceleration of a_k that is normally distributed, with mean 0 and standard deviation σ_a. From Newton's laws of motion we conclude that

$$\mathbf{x}_k = \mathbf{F}\mathbf{x}_{k-1} + \mathbf{G}a_k$$

(note that there is no $\mathbf{B}u$ term since we have no known control inputs) where

$$\mathbf{F} = \begin{bmatrix} 1 & \Delta t \\ 0 & 1 \end{bmatrix}$$

and

$$\mathbf{G} = \begin{bmatrix} \dfrac{\Delta t^2}{2} \\ [6pt]\Delta t \end{bmatrix}$$

so that

$$\mathbf{x}_k = \mathbf{F}\mathbf{x}_{k-1} + \mathbf{w}_k$$

where $\mathbf{w}_k \sim N(0,\mathbf{Q})$ and

$$\mathbf{Q} = \mathbf{G}\mathbf{G}^{\mathrm{T}}\sigma_a^2 = \begin{bmatrix} \dfrac{\Delta t^4}{4} & \dfrac{\Delta t^3}{2} \\ \text{[6}pt\text{]}\dfrac{\Delta t^3}{2} & \Delta t^2 \end{bmatrix}\sigma_a^2.$$

At each time step, a noisy measurement of the true position of the truck is made. Let us suppose the measurement noise v_k is also normally distributed, with mean 0 and standard deviation σ_z.

$$\mathbf{z}_k = \mathbf{H}\mathbf{x}_k + \mathbf{v}_k$$

where

$$\mathbf{H} = \begin{bmatrix} 1 & 0 \end{bmatrix}$$

and

$$\mathbf{R} = \mathrm{E}[\mathbf{v}_k \mathbf{v}_k^{\mathrm{T}}] = \begin{bmatrix} \sigma_z^2 \end{bmatrix}$$

We know the initial starting state of the truck with perfect precision, so we initialize

$$\hat{\mathbf{x}}_{0|0} = \begin{bmatrix} 0 \\ 0 \end{bmatrix}$$

and to tell the filter that we know the exact position and velocity, we give it a zero covariance matrix:

$$\mathbf{P}_{0|0} = \begin{bmatrix} 0 & 0 \\ 0 & 0 \end{bmatrix}$$

If the initial position and velocity are not known perfectly the covariance matrix should be initialized with a suitably large number, say L, on its diagonal.

$$\mathbf{P}_{0|0} = \begin{bmatrix} L & 0 \\ 0 & L \end{bmatrix}$$

The filter will then prefer the information from the first measurements over the information already in the model.

Derivations

Deriving the a Posteriori Estimate Covariance Matrix

Starting with our invariant on the error covariance $P_{k|k}$ as above

$$\mathbf{P}_{k|k} = \text{cov}(\mathbf{x}_k - \hat{\mathbf{x}}_{k|k})$$

substitute in the definition of $\hat{\mathbf{x}}_{k|k}$

$$\mathbf{P}_{k|k} = \text{cov}(\mathbf{x}_k - (\hat{\mathbf{x}}_{k|k-1} + \mathbf{K}_k \tilde{\mathbf{y}}_k))$$

and substitute $\tilde{\mathbf{y}}_k$

$$\mathbf{P}_{k|k} = \text{cov}(\mathbf{x}_k - (\hat{\mathbf{x}}_{k|k-1} + \mathbf{K}_k (\mathbf{z}_k - \mathbf{H}_k \hat{\mathbf{x}}_{k|k-1})))$$

and $\mathbf{z}_k$

$$\mathbf{P}_{k|k} = \text{cov}(\mathbf{x}_k - (\hat{\mathbf{x}}_{k|k-1} + \mathbf{K}_k (\mathbf{H}_k \mathbf{x}_k + \mathbf{v}_k - \mathbf{H}_k \hat{\mathbf{x}}_{k|k-1})))$$

and by collecting the error vectors we get

$$\mathbf{P}_{k|k} = \text{cov}((I - \mathbf{K}_k \mathbf{H}_k)(\mathbf{x}_k - \hat{\mathbf{x}}_{k|k-1}) - \mathbf{K}_k \mathbf{v}_k)$$

Since the measurement error v_k is uncorrelated with the other terms, this becomes

$$\mathbf{P}_{k|k} = \text{cov}((I - \mathbf{K}_k \mathbf{H}_k)(\mathbf{x}_k - \hat{\mathbf{x}}_{k|k-1})) + \text{cov}(\mathbf{K}_k \mathbf{v}_k)$$

by the properties of vector covariance this becomes

$$\mathbf{P}_{k|k} = (I - \mathbf{K}_k \mathbf{H}_k)\text{cov}(\mathbf{x}_k - \hat{\mathbf{x}}_{k|k-1})(I - \mathbf{K}_k \mathbf{H}_k)^{\mathrm{T}} + \mathbf{K}_k \text{cov}(\mathbf{v}_k)\mathbf{K}_k^{\mathrm{T}}$$

which, using our invariant on $P_{k|k-1}$ and the definition of R_k becomes

$$\mathbf{P}_{k|k} = (I - \mathbf{K}_k \mathbf{H}_k)\mathbf{P}_{k|k-1}(I - \mathbf{K}_k \mathbf{H}_k)^{\mathrm{T}} + \mathbf{K}_k \mathbf{R}_k \mathbf{K}_k^{\mathrm{T}}$$

This formula (sometimes known as the "Joseph form" of the covariance update equation) is valid for any value of K_k. It turns out that if K_k is the optimal Kalman gain, this can be simplified further as shown below.

Kalman Gain Derivation

The Kalman filter is a minimum mean-square error estimator. The error in the a posteriori state estimation is

$$\mathbf{x}_k - \hat{\mathbf{x}}_{k|k}$$

We seek to minimize the expected value of the square of the magnitude of this vector, $\mathrm{E}[\| \mathbf{x}_k - \hat{\mathbf{x}}_{k|k} \|^2]$. This is equivalent to minimizing the trace of the a posteriori estimate covariance matrix $\mathbf{P}_{k|k}$. By expanding out the terms in the equation above and collecting, we get:

$$\begin{aligned}\mathbf{P}_{k|k} &= \mathbf{P}_{k|k-1} - \mathbf{K}_k\mathbf{H}_k\mathbf{P}_{k|k-1} - \mathbf{P}_{k|k-1}\mathbf{H}_k^{\mathrm{T}}\mathbf{K}_k^{\mathrm{T}} + \mathbf{K}_k(\mathbf{H}_k\mathbf{P}_{k|k-1}\mathbf{H}_k^{\mathrm{T}} + \mathbf{R}_k)\mathbf{K}_k^{\mathrm{T}} \\ &= \mathbf{P}_{k|k-1} - \mathbf{K}_k\mathbf{H}_k\mathbf{P}_{k|k-1} - \mathbf{P}_{k|k-1}\mathbf{H}_k^{\mathrm{T}}\mathbf{K}_k^{\mathrm{T}} + \mathbf{K}_k\mathbf{S}_k\mathbf{K}_k^{\mathrm{T}}\end{aligned}$$

The trace is minimized when its matrix derivative with respect to the gain matrix is zero. Using the gradient matrix rules and the symmetry of the matrices involved we find that

$$\frac{\partial\, \mathrm{tr}(\mathbf{P}_{k|k})}{\partial\, \mathbf{K}_k} = -2(\mathbf{H}_k\mathbf{P}_{k|k-1})^{\mathrm{T}} + 2\mathbf{K}_k\mathbf{S}_k = 0.$$

Solving this for K_k yields the Kalman gain:

$$\mathbf{K}_k\mathbf{S}_k = (\mathbf{H}_k\mathbf{P}_{k|k-1})^{\mathrm{T}} = \mathbf{P}_{k|k-1}\mathbf{H}_k^{\mathrm{T}}$$

$$\mathbf{K}_k = \mathbf{P}_{k|k-1}\mathbf{H}_k^{\mathrm{T}}\mathbf{S}_k^{-1}$$

This gain, which is known as the optimal Kalman gain, is the one that yields MMSE estimates when used.

Simplification of the a Posteriori Error Covariance Formula

The formula used to calculate the a posteriori error covariance can be simplified when the Kalman gain equals the optimal value derived above. Multiplying both sides of our Kalman gain formula on the right by $S_kK_k^{\mathrm{T}}$, it follows that

$$\mathbf{K}_k\mathbf{S}_k\mathbf{K}_k^{\mathrm{T}} = \mathbf{P}_{k|k-1}\mathbf{H}_k^{\mathrm{T}}\mathbf{K}_k^{\mathrm{T}}$$

Referring back to our expanded formula for the a posteriori error covariance,

$$\mathbf{P}_{k|k} = \mathbf{P}_{k|k-1} - \mathbf{K}_k\mathbf{H}_k\mathbf{P}_{k|k-1} - \mathbf{P}_{k|k-1}\mathbf{H}_k^{\mathrm{T}}\mathbf{K}_k^{\mathrm{T}} + \mathbf{K}_k\mathbf{S}_k\mathbf{K}_k^{\mathrm{T}}$$

we find the last two terms cancel out, giving

$$\mathbf{P}_{k|k} = \mathbf{P}_{k|k-1} - \mathbf{K}_k\mathbf{H}_k\mathbf{P}_{k|k-1} = (I - \mathbf{K}_k\mathbf{H}_k)\mathbf{P}_{k|k-1}.$$

This formula is computationally cheaper and thus nearly always used in practice, but is only correct for the optimal gain.

If arithmetic precision is unusually low causing problems with numerical stability, or if a non-optimal Kalman gain is deliberately used, this simplification cannot be applied; the a posteriori error covariance formula as derived above must be used.

Sensitivity Analysis

The Kalman filtering equations provide an estimate of the state $\hat{\mathbf{x}}_{k|k}$ and its error covariance $\mathbf{P}_{k|k}$ recursively. The estimate and its quality depend on the system parameters and the noise statistics fed as inputs to the estimator. This section analyzes the effect of uncertainties in the statistical inputs to the filter. In the absence of reliable statistics or the true values of noise covariance matrices $\mathbf{Q}_k$ and $\mathbf{R}_k$, the expression

$$\mathbf{P}_{k|k} = (\mathbf{I} - \mathbf{K}_k\mathbf{H}_k)\mathbf{P}_{k|k-1}(\mathbf{I} - \mathbf{K}_k\mathbf{H}_k)^{\mathrm{T}} + \mathbf{K}_k\mathbf{R}_k\mathbf{K}_k^{\mathrm{T}}$$

no longer provides the actual error covariance. In other words, $\mathbf{P}_{k|k} \neq E[(\mathbf{x}_k - \hat{\mathbf{x}}_{k|k})(\mathbf{x}_k - \hat{\mathbf{x}}_{k|k})^{\mathrm{T}}]$. In most real time applications the covariance matrices that are used in designing the Kalman filter are different from the actual noise covariances matrices. This sensitivity analysis describes the behaviour of the estimation error covariance when the noise covariances as well as the system matrices $\mathbf{F}_k$ and $\mathbf{H}_k$ that are fed as inputs to the filter are incorrect.

Thus, the sensitivity analysis describes the robustness (or sensitivity) of the estimator to misspecified statistical and parametric inputs to the estimator.

This discussion is limited to the error sensitivity analysis for the case of statistical uncertainties. Here the actual noise covariances are denoted by $\mathbf{Q}_k^a$ and $\mathbf{R}_k^a$ respectively, whereas the design values used in the estimator are $\mathbf{Q}_k$ and $\mathbf{R}_k$ respectively.

The actual error covariance is denoted by $\mathbf{P}_{k|k}^a$ and $\mathbf{P}_{k|k}$ as computed by the Kalman filter is referred to as the Riccati variable. When $\mathbf{Q}_k \equiv \mathbf{Q}_k^a$ and $\mathbf{R}_k \equiv \mathbf{R}_k^a$, this means that $\mathbf{P}_{k|k} = \mathbf{P}_{k|k}^a$. While computing the actual error covariance using $\mathbf{P}_{k|k}^a = E[(\mathbf{x}_k - \hat{\mathbf{x}}_{k|k})(\mathbf{x}_k - \hat{\mathbf{x}}_{k|k})^{\mathrm{T}}]$, substituting for $\hat{\mathbf{x}}_{k|k}$ and using the fact that $E[\mathbf{w}_k\mathbf{w}_k^{\mathrm{T}}] = \mathbf{Q}_k^a$ and $E[\mathbf{v}_k\mathbf{v}_k^{\mathrm{T}}] = \mathbf{R}_k^a$, results in the following recursive equations for $\mathbf{P}_{k|k}^a$:

$$\mathbf{P}_{k|k-1}^a = \mathbf{F}_k\mathbf{P}_{k-1|k-1}^a\mathbf{F}_k^{\mathrm{T}} + \mathbf{Q}_k^a$$

and

$$\mathbf{P}_{k|k}^a = (\mathbf{I} - \mathbf{K}_k\mathbf{H}_k)\mathbf{P}_{k|k-1}^a(\mathbf{I} - \mathbf{K}_k\mathbf{H}_k)^{\mathrm{T}} + \mathbf{K}_k\mathbf{R}_k^a\mathbf{K}_k^{\mathrm{T}}$$

While computing $\mathbf{P}_{k|k}$, by design the filter implicitly assumes that $E[\mathbf{w}_k\mathbf{w}_k^{\mathrm{T}}] = \mathbf{Q}_k$ and $E[\mathbf{v}_k\mathbf{v}_k^{\mathrm{T}}] = \mathbf{R}_k$. Note that the recursive expressions for $\mathbf{P}_{k|k}^a$ and are $\mathbf{P}_{k|k}$ identical except for the presence of $\mathbf{Q}_k^a$ and $\mathbf{R}_k^a$ in place of the design values $\mathbf{Q}_k$ and $\mathbf{R}_k$ respectively.

Square Root Form

One problem with the Kalman filter is its numerical stability. If the process noise covariance Q_k is small, round-off error often causes a small positive eigenvalue to be computed as a negative number. This renders the numerical representation of the state covariance matrix P indefinite, while its true form is positive-definite.

Positive definite matrices have the property that they have a triangular matrix square root $P = S \cdot S^T$. This can be computed efficiently using the Cholesky factorization algorithm, but more importantly, if the covariance is kept in this form, it can never have a negative diagonal or become asymmetric. An equivalent form, which avoids many of the square root operations required by the matrix square root yet preserves the desirable numerical properties, is the U-D decomposition form, $P = U \cdot D \cdot U^T$, where U is a unit triangular matrix (with unit diagonal), and D is a diagonal matrix.

Between the two, the U-D factorization uses the same amount of storage, and somewhat less computation, and is the most commonly used square root form. (Early literature on the relative efficiency is somewhat misleading, as it assumed that square roots were much more time-consuming than divisions, while on 21-st century computers they are only slightly more expensive.)

Efficient algorithms for the Kalman prediction and update steps in the square root form were developed by G. J. Bierman and C. L. Thornton.

The L·D·LT decomposition of the innovation covariance matrix S_k is the basis for another type of numerically efficient and robust square root filter. The algorithm starts with the LU decomposition as implemented in the Linear Algebra PACKage (LAPACK). These results are further factored into the $L \cdot D \cdot L^T$ structure with methods given by Golub and Van Loan (algorithm 4.1.2) for a symmetric nonsingular matrix. Any singular covariance matrix is pivoted so that the first diagonal partition is nonsingular and well-conditioned. The pivoting algorithm must retain any portion of the innovation covariance matrix directly corresponding to observed state-variables $H_k \cdot x_{k|k-1}$ that are associated with auxiliary observations in y_k. The $L \cdot D \cdot L^T$ square-root filter requires orthogonalization of the observation vector. This may be done with the inverse square-root of the covariance matrix for the auxiliary variables using Method 2 in Higham (2002, p. 263).

Relationship to Recursive Bayesian Estimation

The Kalman filter can be presented as one of the most simple dynamic Bayesian networks. The Kalman filter calculates estimates of the true values of states recursively over time using incoming measurements and a mathematical process model. Similarly, recursive Bayesian estimation calculates estimates of an unknown probability density function (PDF) recursively over time using incoming measurements and a mathematical process model.

In recursive Bayesian estimation, the true state is assumed to be an unobserved Markov process, and the measurements are the observed states of a hidden Markov model (HMM).

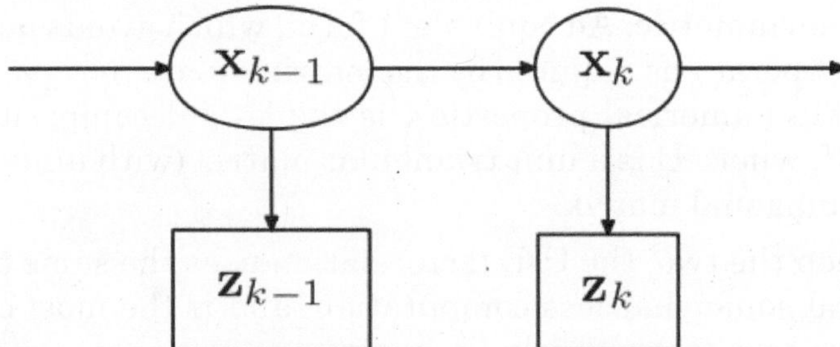

Because of the Markov assumption, the true state is conditionally independent of all earlier states given the immediately previous state.

$$p(\mathbf{x}_k \mid \mathbf{x}_0, \ldots, \mathbf{x}_{k-1}) = p(\mathbf{x}_k \mid \mathbf{x}_{k-1})$$

Similarly the measurement at the k-th timestep is dependent only upon the current state and is conditionally independent of all other states given the current state.

$$p(\mathbf{z}_k \mid \mathbf{x}_0, \ldots, \mathbf{x}_k) = p(\mathbf{z}_k \mid \mathbf{x}_k)$$

Using these assumptions the probability distribution over all states of the hidden Markov model can be written simply as:

$$p(\mathbf{x}_0, \ldots, \mathbf{x}_k, \mathbf{z}_1, \ldots, \mathbf{z}_k) = p(\mathbf{x}_0) \prod_{i=1}^{k} p(\mathbf{z}_i \mid \mathbf{x}_i) p(\mathbf{x}_i \mid \mathbf{x}_{i-1})$$

However, when the Kalman filter is used to estimate the state x, the probability distribution of interest is that associated with the current states conditioned on the measurements up to the current timestep. This is achieved by marginalizing out the previous states and dividing by the probability of the measurement set.

This leads to the predict and update steps of the Kalman filter written probabilistically. The probability distribution associated with the predicted state is the sum (integral) of the products of the probability distribution associated with the transition from the (k − 1)-th timestep to the k-th and the probability distribution associated with the previous state, over all possible x_{k-1}.

$$p(\mathbf{x}_k \mid \mathbf{Z}_{k-1}) = \int p(\mathbf{x}_k \mid \mathbf{x}_{k-1}) p(\mathbf{x}_{k-1} \mid \mathbf{Z}_{k-1}) d\mathbf{x}_{k-1}$$

The measurement set up to time t is

$$\mathbf{Z}_t = \{\mathbf{z}_1, \ldots, \mathbf{z}_t\}$$

The probability distribution of the update is proportional to the product of the measurement likelihood and the predicted state.

$$p(\mathbf{x}_k \mid \mathbf{Z}_k) = \frac{p(\mathbf{z}_k \mid \mathbf{x}_k) p(\mathbf{x}_k \mid \mathbf{Z}_{k-1})}{p(\mathbf{z}_k \mid \mathbf{Z}_{k-1})}$$

The denominator

$$p(\mathbf{z}_k \mid \mathbf{Z}_{k-1}) = \int p(\mathbf{z}_k \mid \mathbf{x}_k) p(\mathbf{x}_k \mid \mathbf{Z}_{k-1}) d\mathbf{x}_k$$

is a normalization term.

The remaining probability density functions are

$$p(\mathbf{x}_k \mid \mathbf{x}_{k-1}) = \mathcal{N}(\mathbf{F}_k \mathbf{x}_{k-1}, \mathbf{Q}_k)$$

$$p(\mathbf{z}_k \mid \mathbf{x}_k) = \mathcal{N}(\mathbf{H}_k \mathbf{x}_k, \mathbf{R}_k)$$

$$p(\mathbf{x}_{k-1} \mid \mathbf{Z}_{k-1}) = \mathcal{N}(\hat{\mathbf{x}}_{k-1}, \mathbf{P}_{k-1})$$

Note that the PDF at the previous timestep is inductively assumed to be the estimated state and covariance. This is justified because, as an optimal estimator, the Kalman filter makes best use of the measurements, therefore the PDF for $\mathbf{x}_k$ given the measurements $\mathbf{Z}_k$ is the Kalman filter estimate.

Information Filter

In the information filter, or inverse covariance filter, the estimated covariance and estimated state are replaced by the information matrix and information vector respectively. These are defined as:

$$\mathbf{Y}_{k|k} = \mathbf{P}_{k|k}^{-1}$$

$$\hat{\mathbf{y}}_{k|k} = \mathbf{P}_{k|k}^{-1}\hat{\mathbf{x}}_{k|k}$$

Similarly the predicted covariance and state have equivalent information forms, defined as:

$$\mathbf{Y}_{k|k-1} = \mathbf{P}_{k|k-1}^{-1}$$

$$\hat{\mathbf{y}}_{k|k-1} = \mathbf{P}_{k|k-1}^{-1}\hat{\mathbf{x}}_{k|k-1}$$

as have the measurement covariance and measurement vector, which are defined as:

$$\mathbf{I}_k = \mathbf{H}_k^{\mathrm{T}}\mathbf{R}_k^{-1}\mathbf{H}_k$$

$$\mathbf{i}_k = \mathbf{H}_k^{\mathrm{T}}\mathbf{R}_k^{-1}\mathbf{z}_k$$

The information update now becomes a trivial sum.

$$\mathbf{Y}_{k|k} = \mathbf{Y}_{k|k-1} + \mathbf{I}_k$$

$$\hat{\mathbf{y}}_{k|k} = \hat{\mathbf{y}}_{k|k-1} + \mathbf{i}_k$$

The main advantage of the information filter is that N measurements can be filtered at each timestep simply by summing their information matrices and vectors.

$$\mathbf{Y}_{k|k} = \mathbf{Y}_{k|k-1} + \sum_{j=1}^{N}\mathbf{I}_{k,j}$$

$$\hat{\mathbf{y}}_{k|k} = \hat{\mathbf{y}}_{k|k-1} + \sum_{j=1}^{N}\mathbf{i}_{k,j}$$

To predict the information filter the information matrix and vector can be converted back to their state space equivalents, or alternatively the information space prediction can be used.

$$\mathbf{M}_k = [\mathbf{F}_k^{-1}]^{\mathrm{T}}\mathbf{Y}_{k-1|k-1}\mathbf{F}_k^{-1}$$

$$\mathbf{C}_k = \mathbf{M}_k[\mathbf{M}_k + \mathbf{Q}_k^{-1}]^{-1}$$

$$\mathbf{L}_k = I - \mathbf{C}_k$$

$$\mathbf{Y}_{k|k-1} = \mathbf{L}_k\mathbf{M}_k\mathbf{L}_k^{\mathrm{T}} + \mathbf{C}_k\mathbf{Q}_k^{-1}\mathbf{C}_k^{\mathrm{T}}$$

$$\hat{\mathbf{y}}_{k|k-1} = \mathbf{L}_k[\mathbf{F}_k^{-1}]^{\mathrm{T}}\hat{\mathbf{y}}_{k-1|k-1}$$

Note that if F and Q are time invariant these values can be cached. Note also that F and Q need to be invertible.

Fixed-Lag Smoother

The optimal fixed-lag smoother provides the optimal estimate of $\hat{\mathbf{x}}_{k-N|k}$ for a given fixed-lag N using the measurements from $\mathbf{z}_1$ to $\mathbf{z}_k$. It can be derived using the previous theory via an augmented state, and the main equation of the filter is the following:

$$\begin{bmatrix} \hat{\mathbf{x}}_{t|t} \\ \hat{\mathbf{x}}_{t-1|t} \\ \vdots \\ \hat{\mathbf{x}}_{t-N+1|t} \end{bmatrix} = \begin{bmatrix} \mathbf{I} \\ 0 \\ \vdots \\ 0 \end{bmatrix} \hat{\mathbf{x}}_{t|t-1} + \begin{bmatrix} 0 & \dots & 0 \\ \mathbf{I} & 0 & \vdots \\ \vdots & \ddots & \vdots \\ 0 & \dots & I \end{bmatrix} \begin{bmatrix} \hat{\mathbf{x}}_{t-1|t-1} \\ \hat{\mathbf{x}}_{t-2|t-1} \\ \vdots \\ \hat{\mathbf{x}}_{t-N+1|t-1} \end{bmatrix} + \begin{bmatrix} \mathbf{K}^{(0)} \\ \mathbf{K}^{(1)} \\ \vdots \\ \mathbf{K}^{(N-1)} \end{bmatrix} \mathbf{y}_{t|t-1}$$

where:

- $\hat{\mathbf{x}}_{t|t-1}$ is estimated via a standard Kalman filter;
- $\mathbf{y}_{t|t-1} = \mathbf{z}(t) - \mathbf{H}\hat{\mathbf{x}}_{t|t-1}$ is the innovation produced considering the estimate of the standard Kalman filter;
- the various $\hat{\mathbf{x}}_{t-i|t}$ with $i = 0,\dots,N$ are new variables, i.e. they do not appear in the standard Kalman filter;
- the gains are computed via the following scheme:

$$\mathbf{K}^{(i)} = \mathbf{P}^{(i)}\mathbf{H}^T\left[\mathbf{HPH}^{\mathrm{T}} + \mathbf{R}\right]^{-1}$$

and

$$\mathbf{P}^{(i)} = \mathbf{P}\left[\left[\mathbf{F} - \mathbf{KH}\right]^T\right]^i$$

where P and Kare the prediction error covariance and the gains of the standard Kalman filter (i.e., $\mathbf{P}_{t|t-1}$).

If the estimation error covariance is defined so that

$$\mathbf{P}_i := E\left[\left(\mathbf{x}_{t-i} - \hat{\mathbf{x}}_{t-i|t}\right)^*\left(\mathbf{x}_{t-i} - \hat{\mathbf{x}}_{t-i|t}\right) \mid z_1 \dots z_t\right],$$

then we have that the improvement on the estimation of $\mathbf{x}_{t-i}$ is given by:

$$\mathbf{P} - \mathbf{P}_i = \sum_{j=0}^{i}\left[\mathbf{P}^{(j)}\mathbf{H}^T\left[\mathbf{HPH}^{\mathrm{T}} + \mathbf{R}\right]^{-1}\mathbf{H}\left(\mathbf{P}^{(i)}\right)^{\mathrm{T}}\right]$$

Fixed-Interval Smoothers

The optimal fixed-interval smoother provides the optimal estimate of $\hat{\mathbf{x}}_{k|n}\ (k < n)$ using the measurements from a fixed interval $\mathbf{z}_1$ to $\mathbf{z}_n$. This is also called "Kalman Smoothing". There are several smoothing algorithms in common use.

Rauch–Tung–Striebel

The Rauch–Tung–Striebel (RTS) smoother is an efficient two-pass algorithm for fixed interval smoothing.

The forward pass is the same as the regular Kalman filter algorithm. These filtered state estimates $\hat{\mathbf{x}}_{k|k}$ and covariances $\mathbf{P}_{k|k}$ are saved for use in the backwards pass.

In the backwards pass, we compute the smoothed state estimates $\hat{\mathbf{x}}_{k|n}$ and covariances $\mathbf{P}_{k|n}$. We start at the last time step and proceed backwards in time using the following recursive equations:

$$\hat{\mathbf{x}}_{k|n} = \hat{\mathbf{x}}_{k|k} + \mathbf{C}_k (\hat{\mathbf{x}}_{k+1|n} - \hat{\mathbf{x}}_{k+1|k})$$

$$\mathbf{P}_{k|n} = \mathbf{P}_{k|k} + \mathbf{C}_k (\mathbf{P}_{k+1|n} - \mathbf{P}_{k+1|k}) \mathbf{C}_k^{\mathrm{T}}$$

where

$$\mathbf{C}_k = \mathbf{P}_{k|k} \mathbf{F}_k^{\mathrm{T}} \mathbf{P}_{k+1|k}^{-1}$$

Modified Bryson–Frazier Smoother

An alternative to the RTS algorithm is the modified Bryson–Frazier (MBF) fixed interval smoother developed by Bierman. This also uses a backward pass that processes data saved from the Kalman filter forward pass. The equations for the backward pass involve the recursive computation of data which are used at each observation time to compute the smoothed state and covariance.

The recursive equations are

$$\tilde{\Lambda}_k = \mathbf{H}_k^T \mathbf{S}_k^{-1} \mathbf{H}_k + \hat{\mathbf{C}}_k^T \hat{\Lambda}_k \hat{\mathbf{C}}_k$$

$$\hat{\Lambda}_{k-1} = \mathbf{F}_k^T \tilde{\Lambda}_k \mathbf{F}_k$$

$$\hat{\Lambda}_n = 0$$

$$\tilde{\lambda}_k = -\mathbf{H}_k^T \mathbf{S}_k^{-1} \mathbf{y}_k + \hat{\mathbf{C}}_k^T \hat{\lambda}_k$$

$$\hat{\lambda}_{k-1} = \mathbf{F}_k^T \tilde{\lambda}_k$$

$$\hat{\lambda}_n = 0$$

where $\mathbf{S}_k$ is the residual covariance and $\widehat{\mathbf{C}}_k = \mathbf{I} - \mathbf{K}_k\mathbf{H}_k$. The smoothed state and covariance can then be found by substitution in the equations

$$\mathbf{P}_{k|n} = \mathbf{P}_{k|k} - \mathbf{P}_{k|k}\hat{\Lambda}_k\mathbf{P}_{k|k}$$

$$\mathbf{x}_{k|n} = \mathbf{x}_{k|k} - \mathbf{P}_{k|k}\hat{\lambda}_k$$

or

$$\mathbf{P}_{k|n} = \mathbf{P}_{k|k-1} - \mathbf{P}_{k|k-1}\tilde{\Lambda}_k\mathbf{P}_{k|k-1}$$

$$\mathbf{x}_{k|n} = \mathbf{x}_{k|k-1} - \mathbf{P}_{k|k-1}\tilde{\lambda}_k.$$

An important advantage of the MBF is that it does not require finding the inverse of the covariance matrix.

Minimum-Variance Smoother

The minimum-variance smoother can attain the best-possible error performance, provided that the models are linear, their parameters and the noise statistics are known precisely. This smoother is a time-varying state-space generalization of the optimal non-causal Wiener filter.

The smoother calculations are done in two passes. The forward calculations involve a one-step-ahead predictor and are given by

$$\hat{\mathbf{x}}_{k+1|k} = (\mathbf{F}_k - \mathbf{K}_k\mathbf{H}_k)\hat{\mathbf{x}}_{k|k-1} + \mathbf{K}_k\mathbf{z}_k$$

$$\alpha_k = -\mathbf{S}_k^{-1/2}\mathbf{H}_k\hat{\mathbf{x}}_{k|k-1} + \mathbf{S}_k^{-1/2}\mathbf{z}_k$$

The above system is known as the inverse Wiener-Hopf factor. The backward recursion is the adjoint of the above forward system. The result of the backward pass β_k may be calculated by operating the forward equations on the time-reversed α_k and time reversing the result. In the case of output estimation, the smoothed estimate is given by

$$\hat{\mathbf{y}}_{k|N} = \mathbf{z}_k - \mathbf{R}_k\beta_k$$

Taking the causal part of this minimum-variance smoother yields

$$\hat{\mathbf{y}}_{k|k} = \mathbf{z}_k - \mathbf{R}_k\mathbf{S}_k^{-1/2}\alpha_k$$

which is identical to the minimum-variance Kalman filter. The above solutions minimize the variance of the output estimation error. Note that the Rauch–Tung–Striebel smoother derivation assumes that the underlying distributions are Gaussian, whereas the minimum-variance solutions do not. Optimal smoothers for state estimation and input estimation can be constructed similarly.

Expectation-maximization algorithms may be employed to calculate approximate maximum likelihood estimates of unknown state-space parameters within minimum-variance filters and smoothers. Often uncertainties remain within problem assumptions. A smoother that accommodates uncertainties can be designed by adding a positive definite term to the Riccati equation.

In cases where the models are nonlinear, step-wise linearizations may be within the minimum-variance filter and smoother recursions (extended Kalman filtering).

Frequency Weighted Kalman Filters

Pioneering research on the perception of sounds at different frequencies was conducted by Fletcher and Munson in the 1930s. Their work led to a standard way of weighting measured sound levels within investigations of industrial noise and hearing loss.

Frequency weightings have since been used within filter and controller designs to manage performance within bands of interest.

Typically, a frequency shaping function is used to weight the average power of the error spectral density in a specified frequency band. Let $\mathbf{y} - \hat{\mathbf{y}}$ denote the output estimation error exhibited by a conventional Kalman filter. Also, let W denote a causal frequency weighting transfer function. The optimum solution which minimizes the variance of $\mathbf{W}(\mathbf{y} - \hat{\mathbf{y}})$ arises by simply constructing $\mathbf{W}^{-1}\hat{\mathbf{y}}$.

The design of $\mathbf{W}$ remains an open question. One way of proceeding is to identify a system which generates the estimation error and setting $\mathbf{W}$ equal to the inverse of that system.

This procedure may be iterated to obtain mean-square error improvement at the cost of increased filter order. The same technique can be applied to smoothers.

Non-Linear Filters

The basic Kalman filter is limited to a linear assumption. More complex systems, however, can be nonlinear. The non-linearity can be associated either with the process model or with the observation model or with both.

Extended Kalman Filter

In the extended Kalman filter (EKF), the state transition and observation models need not be linear functions of the state but may instead be non-linear functions. These functions are of differentiable type.

$$\mathbf{x}_k = f(\mathbf{x}_{k-1}, \mathbf{u}_k) + \mathbf{w}_k$$

$$\mathbf{z}_k = h(\mathbf{x}_k) + \mathbf{v}_k$$

The function f can be used to compute the predicted state from the previous estimate and similarly the function h can be used to compute the predicted measurement from the predicted state. However, f and h cannot be applied to the covariance directly. Instead a matrix of partial derivatives (the Jacobian) is computed.

At each timestep the Jacobian is evaluated with current predicted states. These matrices can be used in the Kalman filter equations. This process essentially linearizes the non-linear function around the current estimate.

Unscented Kalman Filter

When the state transition and observation models—that is, the predict and update functions f and h—are highly non-linear, the extended Kalman filter can give particularly poor performance. This is because the covariance is propagated through linearization of the underlying non-linear model. The unscented Kalman filter (UKF) uses a deterministic sampling technique known as the unscented transform to pick a minimal set of sample points (called sigma points) around the mean. These sigma points are then propagated through the non-linear functions, from which the mean and covariance of the estimate are then recovered.

The result is a filter which more accurately captures the true mean and covariance. (This can be verified using Monte Carlo sampling or through a Taylor series expansion of the posterior statistics.) In addition, this technique removes the requirement to explicitly calculate Jacobians, which for complex functions can be a difficult task in itself (i.e., requiring complicated derivatives if done analytically or being computationally costly if done numerically).

Predict

As with the EKF, the UKF prediction can be used independently from the UKF update, in combination with a linear (or indeed EKF) update, or vice versa.

The estimated state and covariance are augmented with the mean and covariance of the process noise.

$$\mathbf{x}^a_{k-1|k-1} = [\hat{\mathbf{x}}^{\mathrm{T}}_{k-1|k-1} \quad E[\mathbf{w}^{\mathrm{T}}_k]]^{\mathrm{T}}$$

$$\mathbf{P}^a_{k-1|k-1} = \begin{bmatrix} \mathbf{P}_{k-1|k-1} & 0 \\ 0 & \mathbf{Q}_k \end{bmatrix}$$

A set of 2L + 1 sigma points is derived from the augmented state and covariance where L is the dimension of the state.

$$\chi^0_{k-1|k-1} = \mathbf{x}^a_{k-1|k-1}$$

$$\chi^i_{k-1|k-1} = \mathbf{x}^a_{k-1|k-1} + \left(\sqrt{(L+\lambda)\mathbf{P}^a_{k-1|k-1}}\right)_i, \qquad i = 1,\ldots,L$$

$$\chi^i_{k-1|k-1} = \mathbf{x}^a_{k-1|k-1} - \left(\sqrt{(L+\lambda)\mathbf{P}^a_{k-1|k-1}}\right)_{i-L}, \qquad i = L+1,\ldots,2L$$

where

$$\left(\sqrt{(L+\lambda)\mathbf{P}^a_{k-1|k-1}}\right)_i$$

is the ith column of the matrix square root of

$$(L+\lambda)\mathbf{P}^a_{k-1|k-1}$$

using the definition: square root A of matrix B satisfies

$$\mathbf{B} \triangleq \mathbf{A}\mathbf{A}^{\mathrm{T}}.$$

The matrix square root should be calculated using numerically efficient and stable methods such as the Cholesky decomposition.

The sigma points are propagated through the transition function *f*.

$$\chi^i_{k|k-1} = f(\chi^i_{k-1|k-1}) \quad i = 0,\ldots,2L$$

where $f : R^L \rightarrow R^{|\mathbf{x}|}$. The weighted sigma points are recombined to produce the predicted state and covariance.

$$\hat{\mathbf{x}}_{k|k-1} = \sum_{i=0}^{2L} W^i_s \chi^i_{k|k-1}$$

$$\mathbf{P}_{k|k-1} = \sum_{i=0}^{2L} W^i_c\, [\chi^i_{k|k-1} - \hat{\mathbf{x}}_{k|k-1}][\chi^i_{k|k-1} - \hat{\mathbf{x}}_{k|k-1}]^{\mathrm{T}}$$

where the weights for the state and covariance are given by:

$$W_s^0 = \frac{\lambda}{L+\lambda}$$

$$W_c^0 = \frac{\lambda}{L+\lambda} + (1-\alpha^2+\beta)$$

$$W_s^i = W_c^i = \frac{1}{2(L+\lambda)}$$

$$\lambda = \alpha^2(L+\kappa) - L$$

α and k control the spread of the sigma points. β is related to the distribution of x. Normal values are $\alpha = 10^{-3}$, $\kappa = 0$ and $\beta = 2$. If the true distribution of x is Gaussian, $\beta = 2$ is optimal.

The predicted state and covariance are augmented as before, except now with the mean and covariance of the measurement noise.

$$\mathbf{x}_{k|k-1}^a = [\hat{\mathbf{x}}_{k|k-1}^{\mathrm{T}} \quad E[\mathbf{v}_k^{\mathrm{T}}]]^{\mathrm{T}}$$

$$\mathbf{P}_{k|k-1}^a = \begin{bmatrix} \mathbf{P}_{k|k-1} & 0 \\ 0 & \mathbf{R}_k \end{bmatrix}$$

As before, a set of 2L + 1 sigma points is derived from the augmented state and covariance where L is the dimension of the state.

$$\begin{aligned} \chi_{k|k-1}^0 &= \mathbf{x}_{k|k-1}^a \\ \chi_{k|k-1}^i &= \mathbf{x}_{k|k-1}^a + \left(\sqrt{(L+\lambda)\mathbf{P}_{k|k-1}^a}\right)_i, \qquad i = 1,\ldots,L \\ \chi_{k|k-1}^i &= \mathbf{x}_{k|k-1}^a - \left(\sqrt{(L+\lambda)\mathbf{P}_{k|k-1}^a}\right)_{i-L}, \qquad i = L+1,\ldots,2L \end{aligned}$$

Alternatively if the UKF prediction has been used the sigma points themselves can be augmented along the following lines

$$\chi_{k|k-1} := [\chi_{k|k-1}^{\mathrm{T}} \quad E[\mathbf{v}_k^{\mathrm{T}}]]^{\mathrm{T}} \pm \sqrt{(L+\lambda)\mathbf{R}_k^a}$$

where

$$\mathbf{R}_k^a = \begin{bmatrix} 0 & 0 \\ 0 & \mathbf{R}_k \end{bmatrix}$$

The sigma points are projected through the observation function h.

$$\gamma_k^i = h(\chi_{k|k-1}^i) \quad i = 0..2L$$

The weighted sigma points are recombined to produce the predicted measurement and predicted measurement covariance.

$$\hat{\mathbf{z}}_k = \sum_{i=0}^{2L} W_s^i \gamma_k^i$$

$$\mathbf{P}_{z_k z_k} = \sum_{i=0}^{2L} W_c^i [\gamma_k^i - \hat{\mathbf{z}}_k][\gamma_k^i - \hat{\mathbf{z}}_k]^{\mathrm{T}}$$

The state-measurement cross-covariance matrix,

$$\mathbf{P}_{x_k z_k} = \sum_{i=0}^{2L} W_c^i [\chi_{k|k-1}^i - \hat{\mathbf{x}}_{k|k-1}][\gamma_k^i - \hat{\mathbf{z}}_k]^{\mathrm{T}}$$

is used to compute the UKF Kalman gain.

$$K_k = \mathbf{P}_{x_k z_k} \mathbf{P}_{z_k z_k}^{-1}$$

As with the Kalman filter, the updated state is the predicted state plus the innovation weighted by the Kalman gain,

$$\hat{\mathbf{x}}_{k|k} = \hat{\mathbf{x}}_{k|k-1} + K_k(\mathbf{z}_k - \hat{\mathbf{z}}_k)$$

And the updated covariance is the predicted covariance, minus the predicted measurement covariance, weighted by the Kalman gain.

$$\mathbf{P}_{k|k} = \mathbf{P}_{k|k-1} - K_k \mathbf{P}_{z_k z_k} K_k^{\mathrm{T}}$$

Kalman–Bucy Filter

The Kalman–Bucy filter (named after Richard Snowden Bucy) is a continuous time version of the Kalman filter.

It is based on the state space model

$$\frac{d}{dt}\mathbf{x}(t) = \mathbf{F}(t)\mathbf{x}(t) + \mathbf{B}(t)\mathbf{u}(t) + \mathbf{w}(t)$$

$$\mathbf{z}(t) = \mathbf{H}(t)\mathbf{x}(t) + \mathbf{v}(t)$$

where $\mathbf{Q}(t)$ and $\mathbf{R}(t)$ represent the intensities of the two white noise terms $\mathbf{w}(t)$ and $\mathbf{v}(t)$, respectively.

The filter consists of two differential equations, one for the state estimate and one for the covariance:

$$\frac{d}{dt}\hat{\mathbf{x}}(t) = \mathbf{F}(t)\hat{\mathbf{x}}(t) + \mathbf{B}(t)\mathbf{u}(t) + \mathbf{K}(t)(\mathbf{z}(t) - \mathbf{H}(t)\hat{\mathbf{x}}(t))$$

$$\frac{d}{dt}\mathbf{P}(t) = \mathbf{F}(t)\mathbf{P}(t) + \mathbf{P}(t)\mathbf{F}^T(t) + \mathbf{Q}(t) - \mathbf{K}(t)\mathbf{R}(t)\mathbf{K}^T(t)$$

where the Kalman gain is given by

$$\mathbf{K}(t) = \mathbf{P}(t)\mathbf{H}^T(t)\mathbf{R}^{-1}(t)$$

Note that in this expression for $\mathbf{K}(t)$ the covariance of the observation noise $\mathbf{R}(t)$ represents at the same time the covariance of the prediction error (or innovation) $\tilde{\mathbf{y}}(t) = \mathbf{z}(t) - \mathbf{H}(t)\hat{\mathbf{x}}(t)$; these covariances are equal only in the case of continuous time.

The distinction between the prediction and update steps of discrete-time Kalman filtering does not exist in continuous time.

The second differential equation, for the covariance, is an example of a Riccati equation.

Hybrid Kalman Filter

Most physical systems are represented as continuous-time models while discrete-time measurements are frequently taken for state estimation via a digital processor. Therefore, the system model and measurement model are given by

$$\begin{aligned} \mathbf{x}(t) &= \mathbf{F}(t)\mathbf{x}(t) + \mathbf{B}(t)\mathbf{u}(t) + \mathbf{w}(t), \quad \mathbf{w}(t) \sim N\big(\mathbf{0}, \mathbf{Q}(t)\big) \\ \mathbf{z}_k &= \mathbf{H}_k\mathbf{x}_k + \mathbf{v}_k, \quad \mathbf{v}_k \sim N(\mathbf{0}, \mathbf{R}_k) \end{aligned}$$

where

$$\mathbf{x}_k = \mathbf{x}(t_k).$$

Initialize

$$\hat{\mathbf{x}}_{0|0} = E\left[\mathbf{x}(t_0)\right], \mathbf{P}_{0|0} = Var\left[\mathbf{x}(t_0)\right]$$

Predict

$$\hat{\mathbf{x}}(t) = \mathbf{F}(t)\hat{\mathbf{x}}(t) + \mathbf{B}(t)\mathbf{u}(t), \text{ with } \hat{\mathbf{x}}(t_{k-1}) = \hat{\mathbf{x}}_{k-1|k-1}$$

$$\Rightarrow \quad \hat{\mathbf{x}}_{k|k-1} = \hat{\mathbf{x}}(t_k)$$

$$\mathbf{P}(t) = \mathbf{F}(t)\mathbf{P}(t) + \mathbf{P}(t)\mathbf{F}(t)^T + \mathbf{Q}(t), \text{ with } \mathbf{P}(t_{k-1}) = \mathbf{P}_{k-1|k-1}$$

$$\Rightarrow \quad \mathbf{P}_{k|k-1} = \mathbf{P}(t_k)$$

The prediction equations are derived from those of continuous-time Kalman filter without update from measurements, i.e., $\mathbf{K}(t) = 0$. The predicted state and covariance are calculated respectively by solving a set of differential equations with the initial value equal to the estimate at the previous step.

Update

$$\mathbf{K}_k = \mathbf{P}_{k|k-1}\mathbf{H}_k^T\left(\mathbf{H}_k\mathbf{P}_{k|k-1}\mathbf{H}_k^T + \mathbf{R}_k\right)^{-1}$$

$$\hat{\mathbf{x}}_{k|k} = \hat{\mathbf{x}}_{k|k-1} + \mathbf{K}_k(\mathbf{z}_k - \mathbf{H}_k\hat{\mathbf{x}}_{k|k-1})$$

$$\mathbf{P}_{k|k} = (\mathbf{I} - \mathbf{K}_k\mathbf{H}_k)\mathbf{P}_{k|k-1}$$

The update equations are identical to those of the discrete-time Kalman filter.

Variants for the Recovery of Sparse Signals

Recently the traditional Kalman filter has been employed for the recovery of sparse, possibly dynamic, signals from noisy observations. Both works utilize notions from the theory of compressed sensing/sampling, such as the restricted isometry property and related probabilistic recovery arguments, for sequentially estimating the sparse state in intrinsically low-dimensional systems.

Time Domain

Time domain is the analysis of mathematical functions, physical signals or time series of economic or environmental data, with respect to time. In the time domain, the signal or function's value is known for all real numbers, for the case of continuous time, or at various separate instants in the case of discrete time. An oscilloscope is a tool commonly used to visualize real-world signals in the time domain. A time-domain graph shows how a signal changes with time, whereas a frequency-domain graph shows how much of the signal lies within each given frequency band over a range of frequencies.

Origin of Term

The use of the contrasting terms time domain and frequency domain developed in U.S. communication engineering in the late 1940s, with the terms appearing together without definition by 1950. When an analysis uses the second or one of its multiples as a unit of measurement, then it is in the time domain. When analysis concerns the reciprocal units such as Hertz, then it is in the frequency domain.

Chapter 4

State-Space Models

State-space models are models that use state variables to describe a system by a set of first-order differential or difference equations, rather than by one or more nth-order differential or difference equations. State variables x(t) can be reconstructed from the measured input-output data, but are not themselves measured during an experiment.

The state-space model structure is a good choice for quick estimation because it requires you to specify only one input, the model order, n. The model order is an integer equal to the dimension of x(t) and relates to, but is not necessarily equal to, the number of delayed inputs and outputs used in the corresponding linear difference equation.

State Space Representation

In control engineering, a state space representation is a mathematical model of a physical system as a set of input, output and state variables related by first-order differential equations. To abstract from the number of inputs, outputs and states, the variables are expressed as vectors. Additionally, if the dynamical system is linear and time invariant, the differential and algebraic equations may be written in matrix form. The state space representation (also known as the "time-domain approach") provides a convenient and compact way to model and analyze systems with multiple inputs and outputs. With p inputs and q outputs, we would otherwise have to write down $q \times p$ Laplace transforms to encode all the information about a system. Unlike the frequency domain approach, the use of the state space representation is not limited to systems with linear components and zero initial conditions. "State space" refers to the space whose axes are the state variables. The state of the system can be represented as a vector within that space.

State Variables

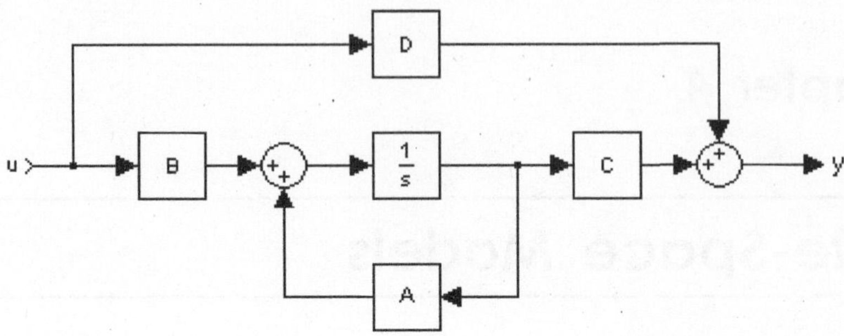

Figure: Block diagram representation of the state space equations

The internal state variables are the smallest possible subset of system variables that can represent the entire state of the system at any given time. The minimum number of state variables required to represent a given system, n, is usually equal to the order of the system's defining differential equation. If the system is represented in transfer function form, the minimum number of state variables is equal to the order of the transfer function's denominator after it has been reduced to a proper fraction.

It is important to understand that converting a state space realisation to a transfer function form may lose some internal information about the system, and may provide a description of a system which is stable, when the state-space realisation is unstable at certain points. In electric circuits, the number of state variables is often, though not always, the same as the number of energy storage elements in the circuit such as capacitors and inductors. The state variables defined must be linearly independent, i.e., no state variable can be written as a linear combination of the other state variables or the system will not be able to be solved.

Linear Systems

The most general state-space representation of a linear system with p inputs, q outputs and n state variables is written in the following form:

$$\dot{\mathbf{x}}(t) = A(t)\mathbf{x}(t) + B(t)\mathbf{u}(t)$$

$$\mathbf{y}(t) = C(t)\mathbf{x}(t) + D(t)\mathbf{u}(t)$$

where:

$\mathbf{x}(\cdot)$ is called the "state vector", $\mathbf{x}(t) \in \mathbb{R}^n$;

$\mathbf{y}(\cdot)$ is called the "output vector", $\mathbf{y}(t) \in \mathbb{R}^q$;

$\mathbf{u}(\cdot)$ is called the "input (or control) vector", $\mathbf{u}(t) \in \mathbb{R}^p$;

$A(\cdot)$ is the "state (or system) matrix", $\dim[A(\cdot)] = n \times n$,

$B(\cdot)$ is the "input matrix", $\dim[B(\cdot)] = n \times p$,

$C(\cdot)$ is the "output matrix", $\dim[C(\cdot)] = q \times n$,

$D(\cdot)$ is the "feedthrough (or feedforward) matrix" (in cases where the system model does not have a direct feedthrough, $D(\cdot)$ is the zero matrix), $\dim[D(\cdot)] = q \times p$,

$$\dot{\mathbf{x}}(t) := \frac{\mathrm{d}}{\mathrm{d}t}\mathbf{x}(t).$$

In this general formulation, all matrices are allowed to be time-variant (i.e. their elements can depend on time); however, in the common LTI case, matrices will be time invariant. The time variable can be continuous (e.g. $t \in \mathbb{R}$) or discrete (e.g. $t \in \mathbb{Z}$). In the latter case, the time variable is usually used instead of . Hybrid systems allow for time domains that have both continuous and discrete parts. Depending on the assumptions taken, the state-space model representation can assume the following forms:

System type	***State-space model***
Continuous time-invariant	$\dot{\mathbf{x}}(t) = A\mathbf{x}(t) + B\mathbf{u}(t)$ $\mathbf{y}(t) = C\mathbf{x}(t) + D\mathbf{u}(t)$
Continuous time-variant	$\mathbf{x}(t) = \mathbf{A}(t)\mathbf{x}(t) + \mathbf{B}(t)\mathbf{u}(t)$ $\mathbf{y}(t) = \mathbf{C}(t)\mathbf{x}(t) + \mathbf{D}(t)\mathbf{u}(t)$
Explicit discrete time-invariant	$\mathbf{x}(k+1) = A\mathbf{x}(k) + B\mathbf{u}(k)$ $\mathbf{y}(k) = C\mathbf{x}(k) + D\mathbf{u}(k)$
Explicit discrete time-variant	$\mathbf{x}(k+1) = \mathbf{A}(k)\mathbf{x}(k) + \mathbf{B}(k)\mathbf{u}(k)$ $\mathbf{y}(k) = \mathbf{C}(k)\mathbf{x}(k) + \mathbf{D}(k)\mathbf{u}(k)$
Laplace domain of continuous time-invariant	$s\mathbf{X}(s) = A\mathbf{X}(s) + B\mathbf{U}(s)$ $\mathbf{Y}(s) = C\mathbf{X}(s) + D\mathbf{U}(s)$
Z-domain of discrete time-invariant	$z\mathbf{X}(z) = A\mathbf{X}(z) + B\mathbf{U}(z)$ $\mathbf{Y}(z) = C\mathbf{X}(z) + D\mathbf{U}(z)$

Example: Continuous-Time LTI Case

Stability and natural response characteristics of a continuous-time LTI system (i.e., linear with matrices that are constant with respect to time) can be studied from the eigenvalues of the matrix A. The stability of a time-invariant state-space model can be determined by looking at the system's transfer function in factored form. It will then look something like this:

$$\mathbf{G}(s) = k\frac{(s-z_1)(s-z_2)(s-z_3)}{(s-p_1)(s-p_2)(s-p_3)(s-p_4)}.$$

The denominator of the transfer function is equal to the characteristic polynomial found by taking the determinant of $sI - A$,

$$\lambda(s) = |sI - A|.$$

The roots of this polynomial (the eigenvalues) are the system transfer function's poles (i.e., the singularities where the transfer function's magnitude is unbounded). These poles can be used to analyze whether the system is asymptotically stable or marginally stable. An alternative approach to determining stability, which does not involve calculating eigenvalues, is to analyze the system's Lyapunov stability.

The zeros found in the numerator of $\mathbf{G}(s)$ can similarly be used to determine whether the system is minimum phase.

The system may still be input–output stable even though it is not internally stable. This may be the case if unstable poles are canceled out by zeros (i.e., if those singularities in the transfer function are removable).

Controllability

State controllability condition implies that it is possible – by admissible inputs – to steer the states from any initial value to any final value within some finite time window. A continuous time-invariant linear state-space model is controllable if and only if

$$\operatorname{rank}\begin{bmatrix} B & AB & A^2B & \dots & A^{n-1}B \end{bmatrix} = n.$$

Where rank is the number of linearly independent rows in a matrix.

Observability

Observability is a measure for how well internal states of a system can be inferred by knowledge of its external outputs. The observability and controllability of a system are mathematical duals (i.e., as controllability provides that an input is available that brings any initial state to any desired final state, observability provides that knowing an output trajectory provides enough information to predict the initial state of the system).

A continuous time-invariant linear state-space model is observable if and only if

$$\text{rank}\begin{bmatrix} C \\ CA \\ \vdots \\ CA^{n-1} \end{bmatrix} = n.$$

Transfer Function

The "transfer function" of a continuous time-invariant linear state-space model can be derived in the following way:

First, taking the Laplace transform of

$$\dot{\mathbf{x}}(t) = A\mathbf{x}(t) + B\mathbf{u}(t)$$

yields

$$s\mathbf{X}(s) = A\mathbf{X}(s) + B\mathbf{U}(s).$$

Next, we simplify for $\mathbf{X}(s)$, giving

$$(s\mathbf{I} - A)\mathbf{X}(s) = B\mathbf{U}(s),$$

and thus

$$\mathbf{X}(s) = (s\mathbf{I} - A)^{-1} B\mathbf{U}(s).$$

Substituting for $\mathbf{X}(s)$ in the output equation

$$\mathbf{Y}(s) = C\mathbf{X}(s) + D\mathbf{U}(s), \text{ giving}$$

$$\mathbf{Y}(s) = C((s\mathbf{I} - A)^{-1} B\mathbf{U}(s)) + D\mathbf{U}(s).$$

Because the transfer function $\mathbf{G}(s)$ is defined as the ratio of the output to the input of a system, we take

$$\mathbf{G}(s) = \mathbf{Y}(s) / \mathbf{U}(s)$$

and substitute the previous expression for $\mathbf{Y}(s)$ with respect to $\mathbf{U}(s)$, giving

$$\mathbf{G}(s) = C(s\mathbf{I} - A)^{-1} B + D.$$

Clearly $\mathbf{G}(s)$ must have q by p dimensionality, and thus has a total of qp elements. So for every input there are transfer functions with one for each output. This is why the state-space representation can easily be the preferred choice for multiple-input, multiple-output (MIMO) systems. The Rosenbrock system matrix provides a bridge between the state-space representation and its transfer function.

Canonical Realisations

Any given transfer function which is strictly proper can easily be transferred into state-space by the following approach (this example is for a 4-dimensional, single-input, single-output system):

Given a transfer function, expand it to reveal all coefficients in both the numerator and denominator. This should result in the following form:

$$\mathbf{G}(s) = \frac{n_1 s^3 + n_2 s^2 + n_3 s + n_4}{s^4 + d_1 s^3 + d_2 s^2 + d_3 s + d_4}.$$

The coefficients can now be inserted directly into the state-space model by the following approach:

$$\dot{\mathbf{x}}(t) = \begin{bmatrix} -d_1 & -d_2 & -d_3 & -d_4 \\ 1 & 0 & 0 & 0 \\ 0 & 1 & 0 & 0 \\ 0 & 0 & 1 & 0 \end{bmatrix} \mathbf{x}(t) + \begin{bmatrix} 1 \\ 0 \\ 0 \\ 0 \end{bmatrix} \mathbf{u}(t)$$

This state-space realisation is called controllable canonical form because the resulting model is guaranteed to be controllable (i.e., because the control enters a chain of integrators, it has the ability to move every state).

The transfer function coefficients can also be used to construct another type of canonical form

$$\dot{\mathbf{x}}(t) = \begin{bmatrix} -d_1 & 1 & 0 & 0 \\ -d_2 & 0 & 1 & 0 \\ -d_3 & 0 & 0 & 1 \\ -d_4 & 0 & 0 & 0 \end{bmatrix} \mathbf{x}(t) + \begin{bmatrix} n_1 \\ n_2 \\ n_3 \\ n_4 \end{bmatrix} \mathbf{u}(t)$$

$$\mathbf{y}(t) = \begin{bmatrix} 1 & 0 & 0 & 0 \end{bmatrix} \mathbf{x}(t).$$

This state-space realisation is called observable canonical form because the resulting model is guaranteed to be observable (i.e., because the output exits from a chain of integrators, every state has an effect on the output).

Proper Transfer Functions

Transfer functions which are only proper (and not strictly proper) can also be realised quite easily. The trick here is to separate the transfer function into two parts: a strictly proper part and a constant.

$$\mathbf{G}(s) = \mathbf{G}_{\text{SP}}(s) + \mathbf{G}(\infty).$$

The strictly proper transfer function can then be transformed into a canonical state space realisation using techniques shown above. The state space realisation of the constant is trivially $\mathbf{y}(t) = \mathbf{G}(\infty)\mathbf{u}(t)$. Together we then get a state space realisation with matrices A, B and C determined by the strictly proper part, and matrix D determined by the constant.

Here is an example to clear things up a bit:

$$\mathbf{G}(s) = \frac{s^2+3s+3}{s^2+2s+1} = \frac{s+2}{s^2+2s+1} + 1$$

which yields the following controllable realisation

$$\dot{\mathbf{x}}(t) = \begin{bmatrix} -2 & -1 \\ 1 & 0 \end{bmatrix} \mathbf{x}(t) + \begin{bmatrix} 1 \\ 0 \end{bmatrix} \mathbf{u}(t)$$

$$\mathbf{y}(t) = \begin{bmatrix} 1 & 2 \end{bmatrix} \mathbf{x}(t) + \begin{bmatrix} 1 \end{bmatrix} \mathbf{u}(t)$$

Notice how the output also depends directly on the input. This is due to the $\mathbf{G}(\infty)$ constant in the transfer function.

Feedback

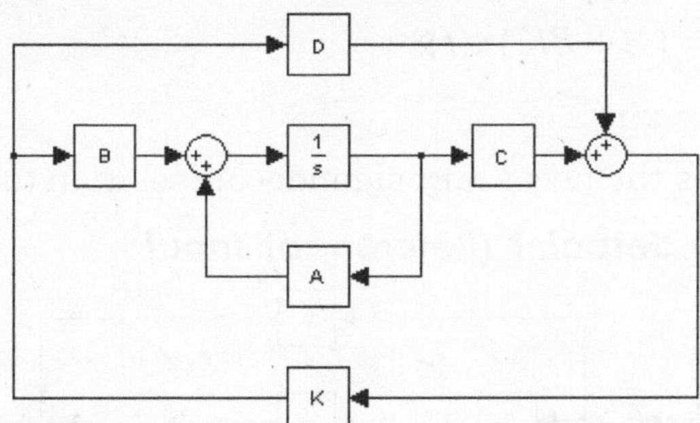

Figure: *Typical state space model with feedback*

A common method for feedback is to multiply the output by a matrix K and setting this as the input to the system: $\mathbf{u}(t) = K\mathbf{y}(t)$. Since the values of K are unrestricted the values can easily be negated for negative feedback. The presence of a negative sign (the common notation) is merely a notational one and its absence has no impact on the end results.

$$\dot{\mathbf{x}}(t) = A\mathbf{x}(t) + B\mathbf{u}(t)$$
$$\mathbf{y}(t) = C\mathbf{x}(t) + D\mathbf{u}(t)$$

becomes

$$\dot{\mathbf{x}}(t) = A\mathbf{x}(t) + BK\mathbf{y}(t)$$
$$\mathbf{y}(t) = C\mathbf{x}(t) + DK\mathbf{y}(t)$$

solving the output equation for $\mathbf{y}(t)$ and substituting in the state equation results in

$$\dot{\mathbf{x}}(t) = \left(A + BK\left(I - DK\right)^{-1} C\right)\mathbf{x}(t)$$

$$\mathbf{y}(t) = \left(I - DK\right)^{-1} C\mathbf{x}(t)$$

The advantage of this is that the eigenvalues of A can be controlled by setting K appropriately through eigendecomposition of $\left(A + BK\left(I - DK\right)^{-1} C\right)$. This assumes that the closed-loop system is controllable or that the unstable eigenvalues of A can be made stable through appropriate choice of *K*.

Example

For a strictly proper system D equals zero. Another fairly common situation is when all states are outputs, i.e. $y = x$, which yields $C = \text{I}$, the Identity matrix. This would then result in the simpler equations

$$\dot{\mathbf{x}}(t) = \left(A + BK\right)\mathbf{x}(t)$$

$$\mathbf{y}(t) = \mathbf{x}(t)$$

This reduces the necessary eigendecomposition to just $A + BK$.

Feedback with Setpoint (Reference) Input

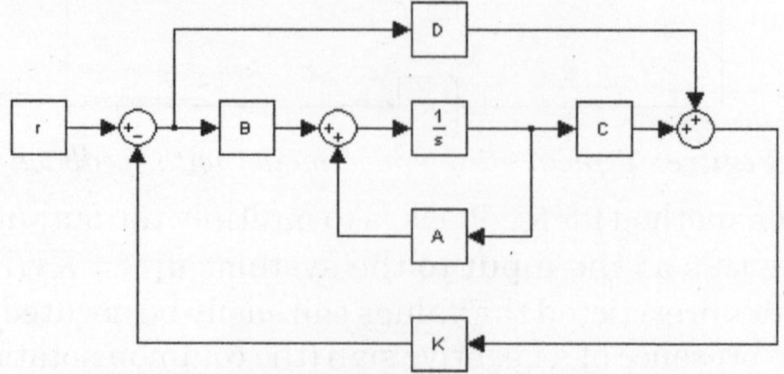

Figure: *Output feedback with set point*

In addition to feedback, an input, $r(t)$, can be added such that

$$\dot{\mathbf{x}}(t) = A\mathbf{x}(t) + B\mathbf{u}(t)$$
$$\mathbf{y}(t) = C\mathbf{x}(t) + D\mathbf{u}(t)$$

becomes

$$\dot{\mathbf{x}}(t) = A\mathbf{x}(t) - BK\mathbf{y}(t) + B\mathbf{r}(t)$$
$$\mathbf{y}(t) = C\mathbf{x}(t) - DK\mathbf{y}(t) + D\mathbf{r}(t)$$

solving the output equation for $\mathbf{y}(t)$ and substituting in the state equation results in

$$\dot{\mathbf{x}}(t) = \left(A - BK(I + DK)^{-1} C\right)\mathbf{x}(t) + B\left(I - K(I + DK)^{-1} D\right)\mathbf{r}(t)$$
$$\mathbf{y}(t) = (I + DK)^{-1} C\mathbf{x}(t) + (I + DK)^{-1} D\mathbf{r}(t)$$

One fairly common simplification to this system is removing D, which reduces the equations to

$$\dot{\mathbf{x}}(t) = (A - BKC)\mathbf{x}(t) + B\mathbf{r}(t)$$
$$\mathbf{y}(t) = C\mathbf{x}(t)$$

Moving Object Example

A classical linear system is that of one-dimensional movement of an object. Newton's laws of motion for an object moving horizontally on a plane and attached to a wall with a spring

$$m\ddot{y}(t) = u(t) - k_1\dot{y}(t) - k_2 y(t)$$

where

- $y(t)$ is position; $\dot{y}(t)$ is velocity; $\ddot{y}(t)$ is acceleration
- $u(t)$ is an applied force
- k_1 is the viscous friction coefficient
- k_2 is the spring constant
- m is the mass of the object

The state equation would then become

$$\begin{bmatrix} \dot{\mathbf{x}}_1(t) \\ \dot{\mathbf{x}}_2(t) \end{bmatrix} = \begin{bmatrix} 0 & 1 \\ -\frac{k_2}{m} & -\frac{k_1}{m} \end{bmatrix} \begin{bmatrix} \mathbf{x}_1(t) \\ \mathbf{x}_2(t) \end{bmatrix} + \begin{bmatrix} 0 \\ \frac{1}{m} \end{bmatrix} \mathbf{u}(t)$$

$$\mathbf{y}(t) = [1 \quad 0]\begin{bmatrix} \mathbf{x}_1(t) \\ \mathbf{x}_2(t) \end{bmatrix}$$

where

- $x_1(t)$ represents the position of the object
- $x_2(t) = \dot{x}_1(t)$ is the velocity of the object
- $\dot{x}_2(t) = \ddot{x}_1(t)$ is the acceleration of the object
- the output $\mathbf{y}(t)$ is the position of the object

The controllability test is then

$$[B \quad AB] = \left[\begin{bmatrix} 0 \\ \dfrac{1}{m} \end{bmatrix} \begin{bmatrix} 0 & 1 \\ -\dfrac{k_2}{m} & -\dfrac{k_1}{m} \end{bmatrix}\begin{bmatrix} 0 \\ \dfrac{1}{m} \end{bmatrix}\right] = \begin{bmatrix} 0 & \dfrac{1}{m} \\ \dfrac{1}{m} & -\dfrac{k_1}{m^2} \end{bmatrix}$$

which has full rank for all k_1 and m.

The observability test is then

$$\begin{bmatrix} C \\ CA \end{bmatrix} = \begin{bmatrix} [1 \quad 0] \\ [1 \quad 0]\begin{bmatrix} 0 & 1 \\ -\dfrac{k_2}{m} & -\dfrac{k_1}{m} \end{bmatrix} \end{bmatrix} = \begin{bmatrix} 1 & 0 \\ 0 & 1 \end{bmatrix}$$

which also has full rank. Therefore, this system is both controllable and observable.

Nonlinear Systems

The more general form of a state space model can be written as two functions.

$$\dot{\mathbf{x}}(t) = \mathbf{f}(t, x(t), u(t))$$

$$\mathbf{y}(t) = \mathbf{h}(t, x(t), u(t))$$

The first is the state equation and the latter is the output equation. If the function $f(\cdot,\cdot,\cdot)$ is a linear combination of states and inputs then the equations can be written in matrix notation like above.

The $u(t)$ argument to the functions can be dropped if the system is unforced (i.e., it has no inputs).

Pendulum Example

A classic nonlinear system is a simple unforced pendulum

$$m\ell^2\ddot{\theta}(t) = -m\ell g\sin\theta(t) - k\ell\dot{\theta}(t)$$

where

- $\theta(t)$ is the angle of the pendulum with respect to the direction of gravity
- m is the mass of the pendulum (pendulum rod's mass is assumed to be zero)
- g is the gravitational acceleration
- k is coefficient of friction at the pivot point
- ℓ is the radius of the pendulum (to the centre of gravity of the mass m)

The state equations are then

$$\dot{x}_1(t) = x_2(t)$$

$$\dot{x}_2(t) = -\frac{g}{\ell}\sin x_1(t) - \frac{k}{m\ell}x_2(t)$$

where

- $x_1(t) = \theta(t)$ is the angle of the pendulum
- $x_2(t) = \dot{x}_1(t)$ is the rotational velocity of the pendulum
- $\dot{x}_2 = \ddot{x}_1$ is the rotational acceleration of the pendulum

Instead, the state equation can be written in the general form

$$\dot{\mathbf{x}}(t) = \begin{pmatrix} \dot{x}_1(t) \\ \dot{x}_2(t) \end{pmatrix} = \mathbf{f}(t, x(t)) = \begin{pmatrix} x_2(t) \\ -\frac{g}{\ell}\sin x_1(t) - \frac{k}{m\ell}x_2(t) \end{pmatrix}.$$

The equilibrium/stationary points of a system are when $\dot{x} = 0$ and so the equilibrium points of a pendulum are those that satisfy

$$\begin{pmatrix} x_1 \\ x_2 \end{pmatrix} = \begin{pmatrix} n\pi \\ 0 \end{pmatrix}$$

for integers n.

Autoregressive–Moving-Average Model

In the statistical analysis of time series, autoregressive–moving-average (ARMA) models provide a parsimonious description of a (weakly) stationary stochastic process in terms of two polynomials, one for the auto-regression and the second for the moving average. The general ARMA model was described in the 1951 thesis of Peter Whittle, Hypothesis testing in time series analysis, and it was popularized in the 1971 book by George E. P. Box and Gwilym Jenkins.

Given a time series of data X_t, the ARMA model is a tool for understanding and, perhaps, predicting future values in this series. The model consists of two parts, an autoregressive (AR) part and a moving average (MA) part. The model is usually then referred to as the ARMA(p,q) model where p is the order of the autoregressive part and q is the order of the moving average part.

The notation AR(p) refers to the autoregressive model of order p. The AR(p) model is written

$$X_t = c + \sum_{i=1}^{p} \varphi_i X_{t-i} + \varepsilon_t.$$

where $\varphi_1, \ldots, \varphi_p$ are parameters, c is a constant, and the random variable ε_t is white noise.

Some constraints are necessary on the values of the parameters of this model in order that the model remains stationary. For example, processes in the AR(1) model with $|\varphi_1| \geq 1$ are not stationary.

Moving-Average Model

The notation MA(*q*) refers to the moving average model of order *q*:

$$X_t = \mu + \varepsilon_t + \sum_{i=1}^{q} \theta_i \varepsilon_{t-i}$$

where the $\theta_1, \ldots, \theta_q$ are the parameters of the model, μ is the expectation of X_t (often assumed to equal 0), and the ε_t, ε_{t-1},... are again, white noise error terms. The moving-average model is essentially a finite impulse response filter with some additional interpretation placed on it.

ARMA Model

The notation ARMA(*p*, *q*) refers to the model with p autoregressive terms and q moving-average terms. This model contains the AR(*p*) and MA(*q*) models,

$$X_t = c + \varepsilon_t + \sum_{i=1}^{p} \varphi_i X_{t-i} + \sum_{i=1}^{q} \theta_i \varepsilon_{t-i}.$$

The general ARMA model was described in the 1951 thesis of Peter Whittle, who used mathematical analysis (Laurent series and Fourier analysis) and statistical inference. ARMA models were popularized by a 1971 book by George E. P. Box and Jenkins, who expounded an iterative (Box–Jenkins) method for choosing and estimating them. This method was useful for low-order polynomials (of degree three or less).

Note about the Error Terms

The error terms ε_t are generally assumed to be independent identically distributed random variables (i.i.d.) sampled from a normal distribution with zero mean: $\varepsilon_t \sim N(0,\sigma^2)$ where σ^2 is the variance. These assumptions may be weakened but doing so will change the properties of the model. In particular, a change to the i.i.d. assumption would make a rather fundamental difference.

Specification in Terms of Lag Operator

In some texts the models will be specified in terms of the lag operator L. In these terms then the AR(p) model is given by

$$\varepsilon_t = \left(1 - \sum_{i=1}^{p} \varphi_i L^i\right) X_t = \varphi(L) X_t$$

where φ represents the polynomial

$$\varphi(L) = 1 - \sum_{i=1}^{p} \varphi_i L^i.$$

The MA(q) model is given by

$$X_t = \left(1 + \sum_{i=1}^{q} \theta_i L^i\right) \varepsilon_t = \theta(L) \varepsilon_t,$$

where θ represents the polynomial

$$\theta(L) = 1 + \sum_{i=1}^{q} \theta_i L^i.$$

Finally, the combined ARMA(p, q) model is given by

$$\left(1 - \sum_{i=1}^{p} \varphi_i L^i\right) X_t = \left(1 + \sum_{i=1}^{q} \theta_i L^i\right) \varepsilon_t,$$

or more concisely,

$$\varphi(L)X_t = \theta(L)\varepsilon_t$$

or

$$\frac{\varphi(L)}{\theta(L)}X_t = \varepsilon_t.$$

Alternative Notation

Some authors, including Box, Jenkins & Reinsel use a different convention for the autoregression coefficients. This allows all the polynomials involving the lag operator to appear in a similar form throughout. Thus the ARMA model would be written as

Fitting Models

ARMA models in general can, after choosing *p* and *q*, be fitted by least squares regression to find the values of the parameters which minimize the error term. It is generally considered good practice to find the smallest values of *p* and *q* which provide an acceptable fit to the data. For a pure AR model the Yule-Walker equations may be used to provide a fit.

Finding appropriate values of *p* and *q* in the ARMA(p,q) model can be facilitated by plotting the partial autocorrelation functions for an estimate of *p*, and likewise using the autocorrelation functions for an estimate of *q*. Further information can be gleaned by considering the same functions for the residuals of a model fitted with an initial selection of *p* and *q*.

Brockwell and Davis recommend using AICc for finding *p* and *q*.

Implementations in Statistics Packages

- In *R*, the arima function (in standard package stats) is documented in ARIMA Modelling of Time Series. Extension packages contain related and extended functionality, e.g., the tseries package includes an arma function, documented in "Fit ARMA Models to Time Series"; the fracdiff package contains fracdiff() for fractionally integrated ARMA processes, etc. The CRAN task view on Time Series contains links to most of these.
- Mathematica has a complete library of time series functions including ARMA.
- MATLAB includes functions such as arma and ar to estimate AR, ARX (autoregressive exogenous), and ARMAX models.

- Statsmodels Python module includes many models and functions for time series analysis, including ARMA. Formerly part of Scikit-learn it is now stand-alone and integrates well with Pandas_(software).
- IMSL Numerical Libraries are libraries of numerical analysis functionality including ARMA and ARIMA procedures implemented in standard programming languages like C, Java, C# .NET, and Fortran.
- gretl can also estimate ARMA model.
- GNU Octave can estimate AR models using functions from the extra package octave-forge.
- Stata includes the function arima which can estimate ARMA and ARIMA models.
- SuanShu is a Java library of numerical methods, including comprehensive statistics packages, in which univariate/ multivariate ARMA, ARIMA, ARMAX, etc. models are implemented in an object-oriented approach. These implementations are documented in "SuanShu, a Java numerical and statistical library".
- SAS has an econometric package, ETS, that estimates ARIMA models.

Applications

ARMA is appropriate when a system is a function of a series of unobserved shocks (the MA part) as well as its own behaviour. For example, stock prices may be shocked by fundamental information as well as exhibiting technical trending and mean-reversion effects due to market participants.

Generalizations

The dependence of X_t on past values and the error terms ε_t is assumed to be linear unless specified otherwise. If the dependence is nonlinear, the model is specifically called a nonlinear moving average (NMA), nonlinear autoregressive (NAR), or nonlinear autoregressive–moving-average (NARMA) model.

Autoregressive–moving-average models can be generalized in other ways. If multiple time series are to be fitted then a vector ARIMA (or VARIMA) model may be fitted. If the time-series in question exhibits long memory then fractional ARIMA (FARIMA, sometimes called

ARFIMA) modelling may be appropriate. If the data is thought to contain seasonal effects, it may be modelled by a SARIMA (seasonal ARIMA) or a periodic ARMA model.

Another generalization is the multiscale autoregressive (MAR) model. A MAR model is indexed by the nodes of a tree, whereas a standard (discrete time) autoregressive model is indexed by integers.

Note that the ARMA model is a univariate model. Extensions for the multivariate case are the Vector Autoregression (VAR) and Vector Autoregression Moving-Average (VARMA).

Autoregressive–Moving-Average Model with Exogenous Inputs Model (ARMAX Model)

The notation ARMAX(p, q, b) refers to the model with p autoregressive terms, q moving average terms and b exogenous inputs terms. This model contains the AR(p) and MA(q) models and a linear combination of the last b terms of a known and external time series d_t. It is given by:

$$X_t = \varepsilon_t + \sum_{i=1}^{p} \varphi_i X_{t-i} + \sum_{i=1}^{q} \theta_i \varepsilon_{t-i} + \sum_{i=0}^{b} \eta_i d_{t-i}.$$

where $\eta_1, \ldots, \eta_b$ are the parameters of the exogenous input d_t.

Statistical packages implement the ARMAX model through the use of "exogenous" or "independent" variables. Care must be taken when interpreting the output of those packages, because the estimated parameters usually refer to the regression:

$$X_t - m_t = \varepsilon_t + \sum_{i=1}^{p} \varphi_i (X_{t-i} - m_{t-i}) + \sum_{i=1}^{q} \theta_i \varepsilon_{t-i}.$$

where m_t incorporates all exogenous (or independent) variables:

$$m_t = c + \sum_{i=0}^{b} \eta_i d_{t-i}.$$

State Space Representation

In control engineering, a state space representation is a mathematical model of a physical system as a set of input, output and state variables related by first-order differential equations. To abstract from the number of inputs, outputs and states, the variables are expressed as vectors.

Additionally, if the dynamical system is linear and time invariant, the differential and algebraic equations may be written in matrix form. The state space representation (also known as the "time-domain approach") provides a convenient and compact way to model and analyze systems with multiple inputs and outputs. With p inputs and q outputs, we would otherwise have to write down $q \times p$ Laplace transforms to encode all the information about a system. Unlike the frequency domain approach, the use of the state space representation is not limited to systems with linear components and zero initial conditions. "State space" refers to the space whose axes are the state variables. The state of the system can be represented as a vector within that space.

State Variables

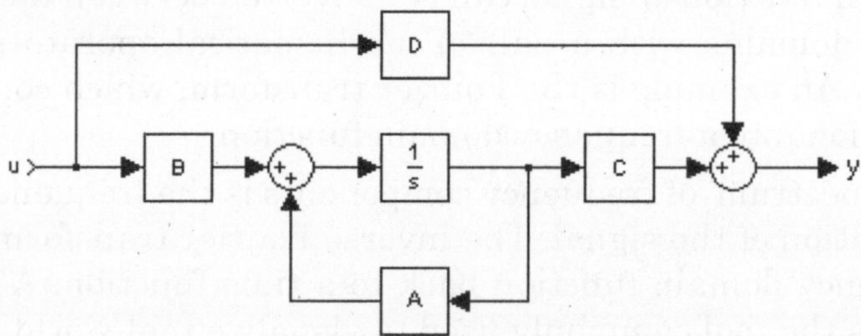

Figure: *Block diagram representation of the state space equations*

The internal state variables are the smallest possible subset of system variables that can represent the entire state of the system at any given time. The minimum number of state variables required to represent a given system, n, is usually equal to the order of the system's defining differential equation. If the system is represented in transfer function form, the minimum number of state variables is equal to the order of the transfer function's denominator after it has been reduced to a proper fraction.

It is important to understand that converting a state space realisation to a transfer function form may lose some internal information about the system, and may provide a description of a system which is stable, when the state-space realisation is unstable at certain points. In electric circuits, the number of state variables is often, though not always, the same as the number of energy storage elements in the circuit such as capacitors and inductors.

The state variables defined must be linearly independent, i.e., no state variable can be written as a linear combination of the other state variables or the system will not be able to be solved.

Frequency Domain

In electronics, control systems engineering, and statistics, the frequency domain refers to the analysis of mathematical functions or signals with respect to frequency, rather than time.

Put simply, a time-domain graph shows how a signal changes over time, whereas a frequency-domain graph shows how much of the signal lies within each given frequency band over a range of frequencies.

A frequency-domain representation can also include information on the phase shift that must be applied to each sinusoid in order to be able to recombine the frequency components to recover the original time signal.

A given function or signal can be converted between the time and frequency domains with a pair of mathematical operators called a transform. An example is the Fourier transform, which converts the time function into a frequency domain function.

The 'spectrum' of frequency components is the frequency domain representation of the signal. The inverse Fourier transform converts the frequency domain function back to a time function. A spectrum analyzer is the tool commonly used to visualize real-world signals in the frequency domain.

Signal processing also allows representations or transforms that result in a joint time-frequency domain, with the instantaneous frequency being a key link between the time domain and the frequency domain.

Magnitude and Phase

In using the Laplace, Z-, or Fourier transforms, the frequency spectrum is complex, describing the magnitude and phase of a signal, or of the response of a system, as a function of frequency. In many applications, phase information is not important.

By discarding the phase information it is possible to simplify the information in a frequency domain representation to generate a frequency spectrum or spectral density. A spectrum analyzer is a device that displays the spectrum, while the time domain frequency can be seen on an oscilloscope.

The power spectral density is a frequency-domain description that can be applied to a large class of signals that are neither periodic nor square-integrable; to have a power spectral density, a signal needs only to be the output of a wide-sense stationary random process.

Different Frequency Domains

Although "the" frequency domain is spoken of in the singular, there are a number of different mathematical transforms which are used to analyze time functions and are referred to as "frequency domain" methods. These are the most common transforms, and the fields in which they are used:

- Fourier series – repetitive signals, oscillating systems
- Fourier transform – nonrepetitive signals, transients
- Laplace transform – electronic circuits and control systems
- Z transform – discrete signals, digital signal processing

More generally, one can speak of the transform domain with respect to any transform. The above transforms can be interpreted as capturing some form of frequency, and hence the transform domain is referred to as a frequency domain.

Discrete Frequency Domain

The Fourier transform of a periodic signal only has energy at a base frequency and its harmonics. Another way of saying this is that a periodic signal can be analyzed using a discrete frequency domain. Dually, a discrete-time signal gives rise to a periodic frequency spectrum.

Combining these two, if we start with a time signal which is both discrete and periodic, we get a frequency spectrum which is both periodic and discrete. This is the usual context for a discrete Fourier transform.

Time Series Analysis in the Frequency Domain

For the detection of smooth signals, e.g. sinusoids, use either ORT/TSA with 'order'=1 or 2 harmonics, SCARGLE/TSA or AOV/TSA with 'order'=3 or 4 bins. The sensitivity of these statistics to sharp signals (such as strongly pulsed variations or light curves of very wide eclipsing binaries) is poor. For the detection of such signals better use ORT/TSA or AOV/TSA with the width of these features matched by the width of the top harmonics or the width of a phase bin, respectively.

The command SINEFIT/TSA serves two purposes: a) least squares estimation of the parameters of a detected signal and b) filtering the data for a given frequency (so-called prewhitening).

The trend removal (zero frequency) constitutes a special case of this filtering. For a pure sinusoid model, the statistic used in SINEFIT/TSA is related to that used in SCARGLE/TSA.

ORT/TSA -

Multiharmonic analysis of variance periodogram: The command computes the analysis of variance (AOV) periodogram for fitting data with a (multiharmonic) Fourier series. The fit of the Fourier series is done by a new efficient algorithm, employing projection onto orthogonal trigonometric polynomials.

The results of the fit are evaluated using the AOV statistics, a powerful method newly adapted for the time series analysis. The model used in this method is the Fourier series of n harmonics. The resolution of the method may be tuned by change of n. Hence it is the method of choice for both smooth and sharp signals.

The AOV statistic is the ratio $S(\nu) = Var_m / Var_r$. The distribution of S for white noise (H_o hypothesis) and order bins is the Fisher-Snedecor distribution F(2n+1,n_o-2n-1). The expected value of the AOV statistics for pure noise is 1 for uncorrelated observations and n_{corr} for observations correlated in groups of size n_{corr}.

SCARGLE/TSA -

Scargle sine model: This command computes Scargle's periodogram for unevenly spaced observations x. The Scargle statistic uses a pure sine model and is a special case of the power spectrum statistic normalized to the variance of the raw data, $| FX^{(m)} |^2 / Var[X^{(o)}]$.

The phase origins of the sinusoids are for each frequency chosen in such a way that the sine and cosine components of become independent. Hence for white noise (H_o hypothesis) S is the ratio of) _ and $\chi^2(n_o)$. For large numbers of observations n_o, numerator and denominator become uncorrelated so that S has a Fisher-Snedecor distribution approaching an exponential distribution in the asymptotic limit: $F(2, n_o - 1) \rightarrow \chi^2(2)/2 = e^{-S}$ for $n \rightarrow \infty$.

We recommend this statistic for larger data sets and for the detection of smooth, nearly sinusoidal signals, since then its test power is large and the statistical properties are known. In particular the expected value is 1. For observations correlated in groups of size n_{corr}, divide the value of the Scargle statistics by n_{corr}. The slow algorithm implemented here is suitable for modest numbers of observations.

SINEFIT/TSA -

Least-squares sinewave fitting: This command fits sine (Fourier) series by nonlinear least squares iterations with simultaneous correction of the frequency. Its main applications are the evaluation of the significance of a detection, parameter estimation, and massaging of data. The values fitted for frequency and Fourier coefficients are displayed on the terminal. For observations correlated in groups of size n_{corr} multiply the errors by $\sqrt{n_{corr}}$. With the latter correction and for purely sinusoidal variations SINEFIT/TSA computes the frequency with an accuracy comparable to the one of the power spectrum. Additionally, the command displays the parameters of the fitted base sinusoid, i.e. of the first Fourier term.

SINEFIT/TSA returns also the table of the residuals $X^{(r)}$ (i.e. of the observations with the fitted oscillation subtracted) in a format suitable for further analysis by any method supported by the TSA package. In this way, the command can be used to perform a CLEAN-like analysis manually by removing individual oscillations one by one in the time domain. Since in most astronomical time series the number of different sinusoids present is quite small, we recommend this manual procedure rather than its automated implementation in frequency space by the CLEAN algorithm.

Alternatively, the command can be used to remove a trend from data. In order to use SINEFIT/TSA for a fixed frequency, specify one iteration only. The corresponding value of may in principle be recovered from the standard deviation $\sigma_o = \sqrt{\chi^2(df)/df}$, where df=$n_{obs}$-$n_{parm}$ and n_{obs} and n_{parm} are the number of observations and the number of Fourier coefficients (including the mean value), respectively. However, the computation of the χ^2 periodogram with SINEFIT/TSA is very cumbersome while the results should correspond exactly to the Scargle periodogram (Scargle, 1982, Lomb, 1976).

AOV/TSA -

Analysis of variance for phase bins: The command computes the analysis of variance (AOV) periodogram for phase folded and binned data. The AOV statistics is a new and powerful method especially suitable for the detection of nonsinusoidal signals (Schwarzenberg-Czerny, 1989). It uses the step function model, i.e. phase binning. Its statistic is $S(\nu) = Var_m / Var_r$.

The distribution of S for white noise (H_0 hypothesis) and $n \equiv$ order bins is the Fisher-Snedecor distribution F(n-1,n_0-n), where n is number of bins. The expected value of the AOV statistics for pure noise is 1 for uncorrelated observations and n_{corr} for observations correlated in groups of size n_{corr}.

Among all statistics named in this chapter, AOV used by ORT/TSA and AOV/TSA is the only one with exactly known statistical properties even for small samples. On large samples, AOV is not less sensitive than other statistics using phase binning, i.e. the step function model: Whittaker & Roberts and PDM. Therefore we recommend the ORT/TSA and AOV/TSA commands for samples of all sizes and particularly for signals with narrow sharp features (pulses, eclipses). If on the average n_{corr} consecutive observations are correlated, divide the value of the periodogram by n_{corr} and use the $F(n-1, n_o/n_{corr}-n)$ distribution. For smooth light curves use low order, e.g. 4 or 3, for optimal sensitivity. For numerous observations and sharp light curves use phase bins of width comparable to that of the narrow features (e.g. pulses, eclipses). Note that phase coverage and consequently quality of the statistics near 0 frequency are notoriously poor for most observations.

Chapter 5

Nonlinear Time Series

Most of the time series models discussed in the previous chapters are linear time series models. Although they remain at the forefront of academic and applied research, it has often been found that simple linear time series models usually leave certain aspects of economic and financial data unexplained. Since economic and financial systems are known to go through both structural and behavioural changes, it is reasonable to assume that different time series models may be required to explain the empirical data at different times. This chapter introduces some popular nonlinear time series models that have been found to be effective at modelling nonlinear behaviour in economic and financial time series data.

To model nonlinear behaviour in economic and financial time series, it seems natural to allow for the existence of different states of the world or regimes and to allow the dynamics to be different in different regimes. This is because simple AR models are arguably the most popular

time series model and are easily estimated using regression methods. By extending AR models to allow for nonlinear behaviour, the resulting nonlinear models are easy to understand and interpret. In addition, this chapter also covers more general Markov switching models using state space representations. The types of models that can be cast into this form are enormous.

In mathematics, a nonlinear system of equations is a set of simultaneous equations in which the unknowns (or the unknown functions in the case of differential equations) appear as variables of a polynomial of degree higher than one or in the argument of a function which is not a polynomial of degree one. In other words, in a nonlinear system of equations, the equation(s) to be solved cannot be written as a linear combination of the unknown variables or functions that appear in it (them). It does not matter if nonlinear known functions appear in the equations. In particular, a differential equation is linear if it is linear in terms of the unknown function and its derivatives, even if nonlinear in terms of the other variables appearing in it.

Typically, the behaviour of a nonlinear system is described by a nonlinear system of equations.

Nonlinear problems are of interest to engineers, physicists and mathematicians and many other scientists because most systems are inherently nonlinear in nature. As nonlinear equations are difficult to solve, nonlinear systems are commonly approximated by linear equations (linearization).

This works well up to some accuracy and some range for the input values, but some interesting phenomena such as chaos and singularities are hidden by linearization. It follows that some aspects of the behaviour of a nonlinear system appear commonly to be chaotic, unpredictable or counterintuitive. Although such chaotic behaviour may resemble random behaviour, it is absolutely not random.

For example, some aspects of the weather are seen to be chaotic, where simple changes in one part of the system produce complex effects throughout. This nonlinearity is one of the reasons why accurate long-term forecasts are impossible with current technology.

Nonlinear Algebraic Equations

Nonlinear algebraic equations, which are also called polynomial equations, are defined by equating polynomials to zero. For example,

$$x^2 + x - 1 = 0.$$

For a single polynomial equation, root-finding algorithms can be used to find solutions to the equation (i.e., sets of values for the variables that satisfy the equation).

However, systems of algebraic equations are more complicated; their study is one motivation for the field of algebraic geometry, a difficult branch of modern mathematics. It is even difficult to decide if a given algebraic system has complex solutions. Nevertheless, in the case of the systems with a finite number of complex solutions, these systems of polynomial equations are now well understood and efficient methods exist for solving them.

Hilbert's Nullstellensatz

Hilbert's Nullstellensatz (German for "theorem of zeros," or more literally, "zero-locus-theorem") is a theorem that establishes a fundamental relationship between geometry and algebra. This relationship is the basis of algebraic geometry, an important branch of mathematics. It relates algebraic sets to ideals in polynomial rings over algebraically closed fields. This relationship was discovered by David Hilbert who proved the Nullstellensatz and several other important related theorems named after him (like Hilbert's basis theorem).

Formulation

Let k be a field (such as the rational numbers) and K be an algebraically closed field extension (such as the complex numbers), consider the polynomial ring $k[X_1,X_2,\ldots, X_n]$ and let I be an ideal in this ring. The algebraic set V(I) defined by this ideal consists of all n-tuples $x = (x_1,\ldots,x_n)$ in K^n such that $f(x) = 0$ for all f in I. Hilbert's Nullstellensatz states that if p is some polynomial in $k[X_1,X_2,\ldots, X_n]$ that vanishes on the algebraic set V(I), i.e. $p(x) = 0$ for all x in V(I), then there exists a natural number r such that p^r is in I.

An immediate corollary is the "weak Nullstellensatz": The ideal I in $k[X_1,X_2,\ldots, X_n]$ contains 1 if and only if the polynomials in I do not have any common zeros in K^n. It may also be formulated as follows: if I is a proper ideal in $k[X_1,X_2,\ldots, X_n]$, then V(I) cannot be empty, i.e. there exists a common zero for all the polynomials in the ideal in every algebraically closed extension of k. This is the reason for the name of the theorem, which can be proved easily from the 'weak' form using the Rabinowitsch trick. The assumption of considering common zeros in an algebraically closed field is essential here; for example, the elements of the proper ideal $(X^2 + 1)$ in R[X] do not have a common zero in R. With the notation common in algebraic geometry, the Nullstellensatz can also be formulated as

$$I(V(J)) = \sqrt{J}$$

for every ideal J. Here, $\sqrt{J}$ denotes the radical of J and I(U) is the ideal of all polynomials that vanish on the set U.

In this way, we obtain an order-reversing bijective correspondence between the algebraic sets in K^n and the radical ideals of $K[X_1,X_2,\ldots, X_n]$. In fact, more generally, one has a Galois connection between subsets of the space and subsets of the algebra, where "Zariski closure" and "radical of the ideal generated" are the closure operators.

As a particular example, consider a point $P=(a_1,\cdots,a_n)\in K^n$. Then $I(P)=(X_1-a_1,\cdots,X_n-a_n)$. More generally,

$$\sqrt{I}=\bigcap_{P\in V(I)}(X_1-a_1,\cdots,X_n-a_n),\quad P=(a_1,\cdots,a_n).$$

As another example, an algebraic subset W in K^n is irreducible (in the Zariski topology) if and only if $I(W)$ is a prime ideal.

Effective Nullstellensatz

In all of its variants, Hilbert's Nullstellensatz asserts that some polynomial g belongs or not to an ideal generated, say, by f_1, ..., f_k; we have $g = f^r$ in the strong version, g = 1 in the weak form. This means the existence or the non existence of polynomials g_1, ..., g_k such that $g = f_1g_1 + ... + f_kg_k$ The usual proofs of the Nullstellensatz are non effective in the sense that they do not give any way to compute the g_i.

It is thus a rather natural question to ask if there is an effective way to compute the g_i (and the exponent r in the strong form) or to prove that they do not exist. To solve this problem, it suffices to provide an upper bound on the total degree of the g_i: such a bound reduces the problem to a finite system of linear equations that may be solved by usual linear algebra techniques. Any such upper bound is called an effective Nullstellensatz.

A related problem is the ideal membership problem, which consists in testing if a polynomial belongs to an ideal. For this problem also, a solution is provided by an upper bound on the degree of the g_i. A general solution of the ideal membership problem provides an effective Nullstellensatz, at least for the weak form.

In 1925, Grete Hermann gave an upper bound for ideal membership problem that is doubly exponential in the number of variables. In 1982 Mayr and Meyer gave an example where the g_i have a degree that is at least double exponential, showing that every general upper bound for the ideal membership problem is doubly exponential in the number of variables.

Until 1987, nobody had the idea that effective Nullstellensatz was easier than ideal membership, when Brownawell gave an upperbound for the effective Nullstellensatz that is simply exponential in the number of variables. Brownawell proof uses calculus techniques and thus is valid only in characteristic 0. Soon after, in 1988, János Kollár gave a purely algebraic proof valid in any characteristic, leading to a better bound.

In the case of the weak Nullstellensatz, Kollár's bound is the following:

Let $f_1, ..., f_s$ be polynomials in $n \geq 2$ variables, of total degree $d_1 \geq ... \geq d_s$. If there exist polynomials g_i such that $f_1 g_1 + ... + f_s g_s = 1$, then they can be chosen such that

$$\deg(f_i g_i) \leq \max(d_s, 3) \prod_{j=1}^{\min(n,s)-1} \max(d_j, 3).$$

This bound is optimal if all the degrees are greater than 2.

If d is the maximum of the degrees of the f_i, this bound may be simplified to

$$\max(3, d)^{\min(n,s)}.$$

Kollár's result has been improved by several authors. M. Sombra has provided the best improvement, up to date, giving the bound

$$\deg(f_i g_i) \leq 2d_s \prod_{j=1}^{\min(n,s)-1} d_j.$$

His bound is better than Kollár's as soon as at least two of the degrees that are involved are lower than 3.

Projective Nullstellensatz

We can formulate a certain correspondence between homogeneous ideals of polynomials and algebraic subsets of a projective space, called the projective Nullstellensatz, that is analogous to the affine one. To do that, we introduce some notations. Let $R = k[t_0,...,t_n]$. The homogeneous ideal $R_+ = \bigoplus_{d \geq 1} R_d$ is called the maximal homogeneous ideal. As in the affine case, we let: for a subset $S \subseteq \mathbb{P}^n$ and a homogeneous ideal I of R,

$$\mathrm{I}_{\mathbb{P}^n}(S) = \{f \in R_+ \mid f = 0 \text{ on } S\},$$

$$\mathrm{V}_{\mathbb{P}^n}(I) = \{x \in \mathbb{P}^n \mid f(x) = 0 \text{ for all } f \in I\}.$$

By $f = 0$ on S we mean: for every homogeneous coordinates $(a_0 : \cdots : a_n)$ of a point of S we have $f(a_0,...,a_n) = 0$.

This implies that the homogeneous components of f are also zero on S and thus that $\mathrm{I}_{\mathbb{P}^n}(S)$ is a homogeneous ideal. Equivalently, $\mathrm{I}_{\mathbb{P}^n}(S)$ is the homogeneous ideal generated by homogeneous polynomials f that vanish on S. Now, for any homogeneous ideal $I \subseteq R_+$, by the usual Nullstellensatz, we have:

$$\sqrt{I} = \mathrm{I}_{\mathbb{P}^n}(\mathrm{V}_{\mathbb{P}^n}(I)),$$

and so, like in the affine case, we have:

There exists an order-reversing one-to-one correspondence between proper homogeneous radical ideals $\mathbb{P}^n$ of R and subsets of of $\mathrm{V}_{\mathbb{P}^n}(I)$. the form The correspondence is given by $\mathrm{I}_{\mathbb{P}^n}$ and $\mathrm{V}_{\mathbb{P}^n}$.

Nonlinear Recurrence Relations

A nonlinear recurrence relation defines successive terms of a sequence as a nonlinear function of preceding terms. Examples of nonlinear recurrence relations are the logistic map and the relations that define the various Hofstadter sequences. Nonlinear discrete models that represent a wide class of nonlinear recurrence relationships include the NARMAX (Nonlinear Autoregressive

Moving Average with eXogenous inputs) model and the related nonlinear system identification and analysis procedures. These approaches can be used to study a wide class of complex nonlinear behaviours in the time, frequency, and spatio-temporal domains.

Nonlinear Differential Equations

A system of differential equations is said to be nonlinear if it is not a linear system. Problems involving nonlinear differential equations are extremely diverse, and methods of solution or analysis are problem dependent. Examples of nonlinear differential equations are the Navier–Stokes equations in fluid dynamics and the Lotka–Volterra equations in biology.

One of the greatest difficulties of nonlinear problems is that it is not generally possible to combine known solutions into new solutions. In linear problems, for example, a family of linearly independent solutions can be used to construct general solutions through the superposition principle.

A good example of this is one-dimensional heat transport with Dirichlet boundary conditions, the solution of which can be written as a time-dependent linear combination of sinusoids of differing frequencies; this makes solutions very flexible. It is often possible to find several very specific solutions to nonlinear equations, however the lack of a superposition principle prevents the construction of new solutions.

Ordinary Differential Equations

First order ordinary differential equations are often exactly solvable by separation of variables, especially for autonomous equations. For example, the nonlinear equation

$$\frac{du}{dx} = -u^2$$

has $u = \frac{1}{x+C}$ as a general solution (and also $u = 0$ as a particular solution, corresponding to the limit of the general solution when C tends to the infinity). The equation is nonlinear because it may be written as

$$\frac{du}{dx} + u^2 = 0$$

and the left-hand side of the equation is not a linear function of u and its derivatives. Note that if the u^2 term were replaced with u, the problem would be linear (the exponential decay problem).

Second and higher order ordinary differential equations (more generally, systems of nonlinear equations) rarely yield closed form solutions, though implicit solutions and solutions involving nonelementary integrals are encountered.

Common methods for the qualitative analysis of nonlinear ordinary differential equations include:

- Examination of any conserved quantities, especially in Hamiltonian systems.
- Examination of dissipative quantities analogous to conserved quantities.
- Linearization via Taylor expansion.
- Change of variables into something easier to study.
- Bifurcation theory.
- Perturbation methods (can be applied to algebraic equations too).

Partial Differential Equations

The most common basic approach to studying nonlinear partial differential equations is to change the variables (or otherwise transform the problem) so that the resulting problem is simpler (possibly even linear). Sometimes, the equation may be transformed into one or more ordinary differential equations, as seen in separation of variables, which is always useful whether or not the resulting ordinary differential equation(s) is solvable.

Another common (though less mathematic) tactic, often seen in fluid and heat mechanics, is to use scale analysis to simplify a general, natural equation in a certain specific boundary value problem.

For example, the (very) nonlinear Navier-Stokes equations can be simplified into one linear partial differential equation in the case of transient, laminar, one dimensional flow in a circular pipe; the scale analysis provides conditions under which the flow is laminar and one dimensional and also yields the simplified equation.

Other methods include examining the characteristics and using the methods outlined above for ordinary differential equations.

Pendula

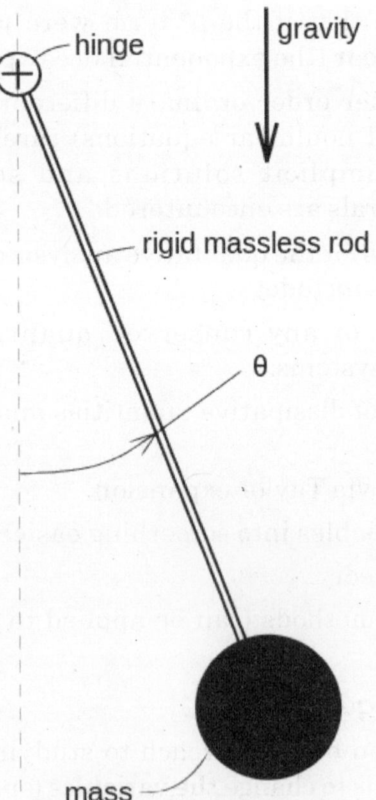

***Figure:** Illustration of a pendulum*

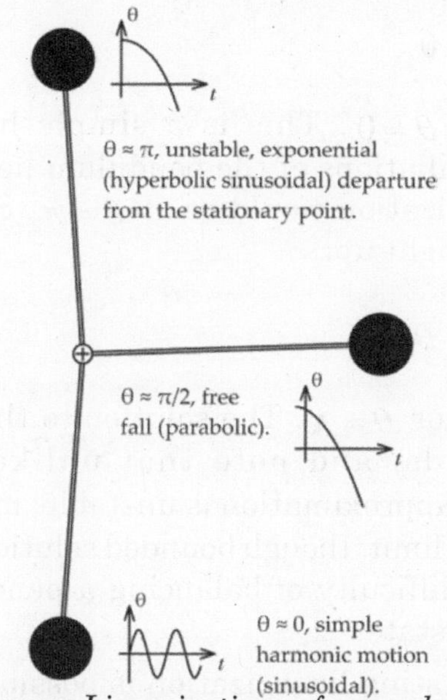

***Figure:** Linearizations of a pendulum*

A classic, extensively studied nonlinear problem is the dynamics of a pendulum under influence of gravity. Using Lagrangian mechanics, it may be shown that the motion of a pendulum can be described by the dimensionless nonlinear equation

$$\frac{d^2\theta}{dt^2} + \sin(\theta) = 0$$

where gravity points "downwards" and θ is the angle the pendulum forms with its rest position, as shown in the figure at right. One approach to "solving" this equation is to use $d\theta / dt$ as an integrating factor, which would eventually yield

$$\int \frac{d\theta}{\sqrt{C_0 + 2\cos(\theta)}} = t + C_1$$

which is an implicit solution involving an elliptic integral. This "solution" generally does not have many uses because most of the nature of the solution is hidden in the nonelementary integral (nonelementary even if $C_0 = 0$).

Another way to approach the problem is to linearize any nonlinearities (the sine function term in this case) at the various points of interest through Taylor expansions. For example, the linearization at $\theta = 0$, called the small angle approximation, is

$$\frac{d^2\theta}{dt^2} + \theta = 0$$

since $\sin(\theta) \approx \theta$ for $\theta \approx 0$. This is a simple harmonic oscillator corresponding to oscillations of the pendulum near the bottom of its path. Another linearization would be at $\theta = \pi$, corresponding to the pendulum being straight up:

$$\frac{d^2\theta}{dt^2} + \pi - \theta = 0$$

since $\sin(\theta) \approx \pi - \theta$ for $\theta \approx \pi$. The solution to this problem involves hyperbolic sinusoids, and note that unlike the small angle approximation, this approximation is unstable, meaning that $|\theta|$ will usually grow without limit, though bounded solutions are possible. This corresponds to the difficulty of balancing a pendulum upright, it is literally an unstable state.

One more interesting linearization is possible around $\theta = \pi / 2$, around which $\sin(\theta) \approx 1$:

$$\frac{d^2\theta}{dt^2} + 1 = 0.$$

This corresponds to a free fall problem. A very useful qualitative picture of the pendulum's dynamics may be obtained by piecing together such linearizations, as seen in the figure at right. Other techniques may be used to find (exact) phase portraits and approximate periods.

Types of Nonlinear Behaviours

- Classical chaos – the behaviour of a system cannot be predicted.
- Multistability – alternating between two or more exclusive states.
- Aperiodic oscillations – functions that do not repeat values after some period (otherwise known as chaotic oscillations or chaos).
- Amplitude death – any oscillations present in the system cease due to some kind of interaction with other system or feedback by the same system.
- Solitons – self-reinforcing solitary waves

High Dimensional Time Series

In statistical theory, the field of high-dimensional statistics studies data whose dimension is larger than dimensions considered in classical multivariate analysis. High-dimensional statistics relies on the theory of random vectors. In many applications, the dimension of the data vectors may be larger than the sample size.

Traditionally, statistical inference considers a probability model for a population and considers data that arose as a sample from the population. For many problems, the estimates of the population characteristics ("parameters") can be substantially refined (in theory) as the sample size increases toward infinity. A traditional requirement of estimators is consistency, that is, the convergence to the unknown true value of the parameter.

In 1968, A.N.Kolmogorov proposed another setting of statistical problems and another setting for the asymptotics, in which the dimension of variables p increases along with the sample size n so that the ratio p/n tends to a constant. It was called the "increasing dimension asymptotics" or "the Kolmogorov asymptotics" Kolmogorov's approach makes it possible to isolate many principal terms of error probabilities and of standard measures of the quality of estimators (quality functions) for large p and n.

Recently, researchers are more interested in even larger dimension cases, e.g. $p = O(\exp(n^a))$, where $0 < a < 1$. This field emerges from the need of extracting meaningful information from many different areas.

Mathematical Theory

Extensive mathematical investigations were carried out that resulted in the creation of systematic theory for improved and asymptotically unimprovable versions of multivariate statistical procedures. A special parameter G that is a function of the fourth moments of variables was found having the property that a small value of G produces a number of specifically many-parametric phenomena. For increasing p and n so that p/n tends to a constant and $G \to 0$, the principal terms of rotation invariant functionals occurring in statistics prove to be dependent on only the first two moments of variables. Under n and p tending to infinity, $p/n \to y > 0$, and $G \to 0$, these functionals have vanishing variance and converge to constants that represent the limit value of empirical means and variances. As a consequence, some stable integral relations are produced between functions of parameters and functions of observable variables. They were called "stochastic canonical equations" or "dispersion equations". Using them one can express the principle parts of standard quality functions of regularized multivariate statistical procedures as functions of only observed variables. This provides the possibility of choosing better procedures and finding asymptotically unimprovable solutions.

Modern statistical and econometric studies frequently face high dimensional time series. Examples come from financial, economic, climate data, functional data analysis, medical imaging and pharmacokinetics, among many others. The stylized feature of such data can be unknown dependence between components, time inhomogeneity in mean and in volatility, structural changes. Modelling and forecasting for such dynamic system requires understanding and recovery of its underlying structure.

The structural assumptions for the considered set of time series allow for reducing its dimensionality and complexity. Classical linear time series modelling fail to mimic many essential features of the data and nonlinear models with unknown link functions become more and more popular. An important example is provided by dynamic factors models.

Dimensional Analysis

In engineering and science, dimensional analysis is the analysis of the relationships between different physical quantities by identifying their fundamental dimensions (such as length, mass, time, and electric charge) and units of measure (such as miles vs. kilometers, or pounds vs. kilograms vs. grams) and tracking these dimensions as calculations or comparisons are performed. Converting from one dimensional unit to another is often somewhat complex. Dimensional analysis, or more specifically the factor-label method, also known as the unit-factor method, is a widely used technique for performing such conversions using the rules of algebra.

Any physically meaningful equation (and any inequality and inequation) must have the same dimensions on the left and right sides. Checking this is a common application of performing dimensional analysis. Dimensional analysis is also routinely used as a check on the plausibility of derived equations and computations. It is generally used to categorize types of physical quantities and units based on their relationship to or dependence on other units.

Concrete Numbers and Fundamental Units

Many parameters and measurements in the physical sciences and engineering are expressed as a concrete number - a numerical quantity and a corresponding dimensional unit. Often a quantity is a combination of multiple dimensions; for example, speed is a combination of length and time, e.g. 60 miles per hour or 1.4 km per second. Compound relations with "per" are expressed with division, e.g. 60 mi/1 *h*. Other relations can involve multiplication (often shown with · or implied), exponents (like m^2 for square meters), or combinations thereof.

Sometimes the names of units obscure their compound nature. For example, an ampere is a measure of electrical current, which is fundamentally electrical charge per unit time and is measured in coulombs (a unit of electrical charge) per second, so 1A = 1C/s. One Newton is kg ·m/s^2

A unit of measure which is not compound - it cannot be decomposed into other units - is known as a fundamental unit.

Percentages and Derivatives

Percentages are dimensionless quantities, since they are ratios of two quantities with the same dimensions. In other words, the % sign can be read as "1/100", since 1% = 1/100.

Derivatives with respect to a quantity add the dimensions of the variable one is differentiating with respect to on the denominator. Thus:

- position (x) has units of L (Length);
- derivative of position with respect to time (dx/dt, velocity) has units of L/T – Length from position, Time from the derivative;
- the second derivative (d^2x/dt^2, acceleration) has units of L/T^2.

In economics, one distinguishes between stocks and flows: a stock has units of "units" (say, widgets or dollars), while a flow is a derivative of a stock, and has units of "units/time" (say, dollars/year).

In some contexts, dimensional quantities are expressed as dimensionless quantities or percentages by omitting some dimensions. For example, Debt to GDP ratios are generally expressed as percentages: total debt outstanding (dimension of Currency) divided by annual GDP (dimension of Currency) – but one may argue that in comparing a stock to a flow, annual GDP should have dimensions of Currency/Time (Dollars/Year, for instance), and thus Debt to GDP should have units of years.

Conversion Factor

In dimensional analysis, a ratio which converts one unit of measure into another without changing the quantity is called a conversion factor. For example, kPa and bar are both units of pressure, and 100 kPa = 1 bar. The rules of algebra allow both sides of an equation to be divided by the same expression, so this is equivalent to 100 kPa/1 bar = 1.

Since any quantity can be multiplied by 1 without changing it, the expression "100 kPa/1 bar" can be used to convert from bars to kPa by multiplying it with the quantity to be converted, including units. For example, 5 bar * 100 kPa/1 bar = 500 kPa because 5*100/1=500, and bar/bar cancels out, so 5 bar = 500 kPa.

Dimensional Homogeneity

The most basic rule of dimensional analysis is that of dimensional homogeneity. Only commensurable quantities (quantities with the same dimensions) may be compared, equated, added, or subtracted. However, the dimensions form a "multiplicative group" and consequently:

One may take ratios of incommensurable quantities (quantities with different dimensions), and multiply or divide them.

For example, it makes no sense to ask if 1 hour is more, the same, or less than 1 kilometer, as these have different dimensions, nor to add 1 hour to 1 kilometer. On the other hand, if an object travels 100 km in 2 hours, one may divide these and conclude that the object's average speed was 50 km/h.

The rule implies that in a physically meaningful expression only quantities of the same dimension can be added, subtracted, or compared. For example, if m_{man}, m_{rat} and L_{man} denote, respectively, the mass of some man, the mass of a rat and the length of that man, the dimensionally homogeneous expression $m_{man} + m_{rat}$ is meaningful, but the heterogeneous expression $m_{man} + L_{man}$ is meaningless. However, m_{man}/L^2_{man} is fine. Thus, dimensional analysis may be used as a sanity check of physical equations: the two sides of any equation must be commensurable or have the same dimensions.

Even when two physical quantities have identical dimensions, it may nevertheless be meaningless to compare or add them. For example, although torque and energy share the dimension ML^2/T^2, they are fundamentally different physical quantities.

To compare, add, or subtract quantities with the same dimensions but expressed in different units, the standard procedure is first to convert them all to the same units. For example, to compare 32 metres with 35 yards, use 1 yard = 0.9144 m to convert 35 yards to 32.004 m.

A related principle is that any physical law that accurately describes the real world must be independent of the units used to measure the physical variables. For example, Newton's laws of motion must hold true whether distance is measured in miles or kilometers.

This principle gives rise to the form that conversion factors must take between units that measure the same dimension: multiplication by a simple constant. It also ensures equivalence; for example, if two buildings are the same height in feet, then they must be the same height in meters.

The Factor-Label Method for Converting Units

The factor-label method is the sequential application of conversion factors expressed as fractions and arranged so that any dimensional unit appearing in both the numerator and denominator of any of the fractions can be cancelled out until only the desired set of dimensional units is obtained. For example, 10 miles per hour can be converted to meters per second by using a sequence of conversion factors as shown below:

$$\frac{10\ \cancel{\text{mile}}}{1\ \cancel{\text{hour}}} \times \frac{1609\ \text{meter}}{1\ \cancel{\text{mile}}} \times \frac{1\ \cancel{\text{hour}}}{3600\ \text{second}} = 4.47\ \frac{\text{meter}}{\text{second}}.$$

It can be seen that each conversion factor is equivalent to the value of one. For example, starting with 1 mile = 1609 meters and dividing both sides of the equation by 1 mile yields 1 mile / 1 mile = 1609 meters / 1 mile, which when simplified yields 1 = 1609 meters / 1 mile.

So, when the units mile and hour are cancelled out and the arithmetic is done, 10 miles per hour converts to 4.47 meters per second.

As a more complex example, the concentration of nitrogen oxides (i.e., NOx) in the flue gas from an industrial furnace can be converted to a mass flow rate expressed in grams per hour (i.e., g/h) of NOx by using the following information as shown below:

NOx concentration

= 10 parts per million by volume = 10 ppmv = 10 volumes/10^6 volumes

NOx molar mass

= 46 kg/kgmol (sometimes also expressed as 46 kg/kmol)

Flow rate of flue gas

= 20 cubic meters per minute = 20 m^3/min

The flue gas exits the furnace at 0 °C temperature and 101.325 kPa absolute pressure.

The molar volume of a gas at 0 °C temperature and 101.325 kPa is 22.414 m^3/kgmol.

$$\frac{10\ \cancel{\text{m}^3\ \text{NOx}}}{10^6\ \cancel{\text{m}^3\ \text{gas}}} \times \frac{20\ \cancel{\text{m}^3\ \text{gas}}}{1\ \cancel{\text{minute}}} \times \frac{60\ \cancel{\text{minute}}}{1\ \text{hour}} \times \frac{1\ \cancel{\text{kgmol NOx}}}{22.414\ \cancel{\text{m}^3\ \text{gas}}} \times \frac{46\ \cancel{\text{kg}}\ \text{NOx}}{1\ \cancel{\text{kgmol NOx}}} \times \frac{1000\ \text{g}}{1\ \cancel{\text{kg}}} = 24.63\ \frac{\text{g NOx}}{\text{hour}}$$

After cancelling out any dimensional units that appear both in the numerators and denominators of the fractions in the above equation, the NOx concentration of 10 ppm_v converts to mass flow rate of 24.63 grams per hour.

Checking Equations that Involve Dimensions

The factor-label method can also be used on any mathematical equation to check whether or not the dimensional units on the left hand side of the equation are the same as the dimensional units on the right hand side of the equation. Having the same units on both sides of an equation does not guarantee that the equation is correct, but having different units on the two sides of an equation does guarantee that the equation is wrong.

For example, check the Universal Gas Law equation of $P \cdot V = n \cdot R \cdot T$, when:

- the pressure P is in pascals (Pa)
- the volume V is in cubic meters (m^3)
- the amount of substance n is in moles (mol)
- the universal gas law constant R is 8.3145 $Pa \cdot m^3/(mol \cdot K)$
- the temperature T is in kelvins (K)

$$\text{Pa m}^3 = \frac{\cancel{\text{mol}}}{1} \times \frac{\text{Pa m}^3}{\cancel{\text{mol}}\,\cancel{\text{K}}} \times \frac{\cancel{\text{K}}}{1}$$

As can be seen, when the dimensional units appearing in the numerator and denominator of the equation's right hand side are cancelled out, both sides of the equation have the same dimensional units.

Limitations

The factor-label method can convert only unit quantities for which the units are in a linear relationship intersecting at 0. Most units fit this paradigm. An example for which it cannot be used is the conversion between degrees Celsius and kelvins (or Fahrenheit). Between degrees Celsius and kelvins, there is a constant difference rather than a constant ratio, while between Celsius and Fahrenheit there is neither a constant difference nor a constant ratio. There is however an affine transform ($ax+b$), (rather than a linear transform (ax)) between them.

For instance, the freezing point of water is 0° in Celsius and 32° in Fahrenheit, and a 5° change in Celsius corresponds to a 9° change in Fahrenheit. Thus, to convert from Fahrenheit to Celsius one subtracts 32° (displacement from one point), multiplies by 5 and divides by 9 (scales by the ratio of units), and adds 0 (displacement from new point). Reversing this yields the formula for Celsius; one could have started with the equivalence between 100° Celsius and 212° Fahrenheit, though this would yield the same formula at the end.

Hence, to convert Fahrenheit to Celsius, enter the Fahrenheit value into this formula:

$$°C = (°F - 32°) \div 1.8$$

Note that dividing by 1.8 is the same as multiplying by 5 and dividing by 9.

And, to convert Celsius to Fahrenheit, enter the Celsius value into this formula:

$$°F = 1.8(°C) + 32°$$

Note that multiplying by 1.8 is the same as multiplying by 9 and dividing by 5.

Applications

Dimensional analysis is most often used in physics and chemistry- and in the mathematics thereof- but finds some applications outside of those fields as well.

Mathematics

A simple application of dimensional analysis to mathematics is in computing the form of the volume of an n-ball (the solid ball in n-dimensions), or the area of its surface, the n-sphere: being an n-dimensional figure, the volume scales as x^n, while the surface area, being $(n-1)$-dimensional, scales as x^{n-1}. Thus the volume of the n-ball in terms of the radius is $C_n r^n$, for some constant C_n. Determining the constant takes more involved mathematics, but the form can be deduced and checked by dimensional analysis alone.

Finance, Economics, and Accounting

In finance, economics, and accounting, dimensional analysis is most commonly referred to in terms of the distinction between stocks and flows. More generally, dimensional analysis is used in interpreting various financial ratios, economics ratios, and accounting ratios.

- For example, the P/E ratio has dimensions of time (units of years), and can be interpreted as "years of earnings to earn the price paid."
- In economics, debt-to-GDP ratio also has units of years (debt has units of currency, GDP has units of currency/year).

- More surprisingly, bond duration also has units of years, which can be shown by dimensional analysis, but takes some financial intuition to understand.
- Velocity of money has units of 1/Years (GDP/Money supply has units of Currency/Year over Currency): how often a unit of currency circulates per year.
- Interest rates are often expressed as a percentage, but more properly percent per annum, which has dimensions of 1/Years.

Critics of mainstream economics, notably including adherents of Austrian economics, have claimed that it lacks dimensional consistency.

Fluid Mechanics

Common dimensionless groups in fluid mechanics include:

- Reynolds number(Re) generally important in all types of fluid problems.

  Re=ρVL/μ

- Froude number(Fr) modelling flow with a free surface.

  Fr=V/sqrt(gL)

- Euler number(Eu) used in problems in which pressure is of interest.

  P/(ρV^2)

History

James Clerk Maxwell played a major role in establishing modern use of dimensional analysis by distinguishing mass, length, and time as fundamental units, while referring to other units as derived.

The 19th-century French mathematician Joseph Fourier made important contributions based on the idea that physical laws like F = ma should be independent of the units employed to measure the physical variables. This led to the conclusion that meaningful laws must be homogeneous equations in their various units of measurement, a result which was eventually formalized in the Buckingham π theorem.

Dimensional analysis is also used to derive relationships between the physical quantities that are involved in a particular phenomenon that one wishes to understand and characterize.

It was used for the first time (Pesic 2005) in this way in 1872 by Lord Rayleigh, who was trying to understand why the sky is blue. Rayleigh first published the technique in his book "theory of sound" from 1877.

Mathematical Examples

The Buckingham π theorem describes how every physically meaningful equation involving n variables can be equivalently rewritten as an equation of *n–m* dimensionless parameters, where m is the rank of the dimensional matrix. Furthermore, and most importantly, it provides a method for computing these dimensionless parameters from the given variables.

A dimensional equation can have the dimensions reduced or eliminated through nondimensionalization, which begins with dimensional analysis, and involves scaling quantities by characteristic units of a system or natural units of nature. This gives insight into the fundamental properties of the system, as illustrated in the examples below.

Definition

The dimension of a physical quantity can be expressed as a product of the basic physical dimensions mass, length, time, electric charge, and absolute temperature, represented by sans-serif symbols M, L, T, Q, and Θ, respectively, each raised to a rational power.

The SI standard recommends the usage of the following dimensions and corresponding symbols: mass (M), length (L), time (T), electrical current (I), absolute temperature Θ, amount of substance (N) and luminous intensity (J).

The term dimension is more abstract than scale unit: mass is a dimension, while kilograms are a scale unit (choice of standard) in the mass dimension.

As examples, the dimension of the physical quantity speed is length/time (L/T or LT^{-1}), and the dimension of the physical quantity force is "mass × acceleration" or "mass×(length/time)/time" (ML/T^2 or MLT^{-2}). In principle, other dimensions of physical quantity could be defined as "fundamental" (such as momentum or energy or electric current) in lieu of some of those shown above. Most physicists do not recognise temperature, as a fundamental dimension of physical quantity since it essentially expresses the energy per particle per degree of freedom, which can be expressed in terms of energy (or mass, length, and time). Still others do not recognise electric current, I, as a separate fundamental dimension of physical quantity, since it has been expressed in terms of mass, length, and time in unit systems such as the cgs system. There are also physicists that have cast doubt on the very existence of incompatible fundamental dimensions of physical quantity, although this does not invalidate the usefulness of dimensional analysis.

The unit of a physical quantity and its dimension are related, but not identical concepts. The units of a physical quantity are defined by convention and related to some standard; e.g. length may have units of metres, feet, inches, miles or micrometres; but any length always has a dimension of L, no matter what units of length are chosen to measure it. Two different units of the same physical quantity have conversion factors that relate them. For example, 1 in = 2.54 cm; in this case (2.54 cm/in) is the conversion factor, and is itself dimensionless. Therefore multiplying by that conversion factor does not change a quantity. Dimensional symbols do not have conversion factors.

Mathematical Properties

The dimensions that can be formed from a given collection of basic physical dimensions, such as M, L, and T, form an abelian group: The identity is written as 1; $L^0 = 1$, and the inverse to L is 1/L or L^{-1}. L raised to any rational power p is a member of the group, having an inverse of L^{-p} or $1/L^p$. The operation of the group is multiplication, having the usual rules for handling exponents ($L^n \times L^m = L^{n+m}$).

This group can be described as a vector space over the rational numbers, with for example dimensional symbol $M^iL^jT^k$ corresponding to the vector (i, j, k). When physical measured quantities (be they like-dimensioned or unlike-dimensioned) are multiplied or divided by one other, their dimensional units are likewise multiplied or divided; this corresponds to addition or subtraction in the vector space. When measurable quantities are raised to a rational power, the same is done to the dimensional symbols attached to those quantities; this corresponds to scalar multiplication in the vector space.

A basis for a given vector space of dimensional symbols is called a set of fundamental units or fundamental dimensions, and all other vectors are called derived units. As in any vector space, one may choose different bases, which yields different systems of units (e.g., choosing whether the unit for charge is derived from the unit for current, or vice versa).

The group identity 1, the dimension of dimensionless quantities, corresponds to the origin in this vector space.

The set of units of the physical quantities involved in a problem correspond to a set of vectors (or a matrix). The kernel describes some number (e.g., m) of ways in which these vectors can be combined to produce a zero vector. These correspond to producing (from the measurements) a number of dimensionless quantities, $\{\pi_1,...,\pi_m\}$. (In fact these ways completely span the null subspace of another different space, of powers of the measurements.) Every possible way of multiplying (and exponentiating) together the measured quantities to produce something with the same units as some derived quantity X can be expressed in the general form

$$X = \prod_{i=1}^{m} (\pi_i)^{k_i}.$$

Consequently, every possible commensurate equation for the physics of the system can be rewritten in the form

$$f(\pi_1, \pi_2, ..., \pi_m) = 0.$$

Knowing this restriction can be a powerful tool for obtaining new insight into the system.

Mechanics

In mechanics, the dimension of any physical quantity can be expressed in terms of the fundamental dimensions (or base dimensions) M, L, and T – these form a 3-dimensional vector space. This is not the only possible choice, but it is the one most commonly used. For example, one might choose force, length and mass as the base dimensions (as some have done), with associated dimensions F, L, M; this corresponds to a different basis, and one may convert between these representations by a change of basis. The choice of the base set of dimensions is, thus, partly a convention, resulting in increased utility and familiarity. It is, however, important to note that the choice of the set of dimensions cannot be chosen arbitrarily – it is not just a convention – because the dimensions must form a basis: they must span the space, and be linearly independent.

For example, F, L, M form a set of fundamental dimensions because they form an equivalent basis to M, L, T: the former can be expressed as [$F=ML/T^2$],L,M while the latter can be expressed as M,L,[$T=(ML/F)^{1/2}$].

On the other hand, using length, velocity and time (L, V, T) as base dimensions will not work well (they do not form a set of fundamental dimensions), for two reasons:

- There is no way to obtain mass – or anything derived from it, such as force – without introducing another base dimension (thus these do not span the space).
- Velocity, being derived from length and time (V=L/T), is redundant (the set is not linearly independent).

Other Fields of Physics and Chemistry

Depending on the field of physics, it may be advantageous to choose one or another extended set of dimensional symbols. In electromagnetism, for example, it may be useful to use dimensions of M, L, T, and Q, where Q represents quantity of electric charge. In thermodynamics, the base set of dimensions is often extended to include a dimension for temperature, È.

In chemistry the number of moles of substance (loosely, but not precisely, related to the number of molecules or atoms) is often involved and a dimension for this is used as well. In the interaction of relativistic plasma with strong laser pulses a dimensionless relativistic similarity parameter connected with the symmetry properties of the collisionless Vlasov equation is constructed from the plasma electron and critical densities in addition to the electromagnetic vector potential. The choice of the dimensions or even the number of dimensions to be used in different fields of physics is to some extent arbitrary, but consistency in use and ease of communications are very important.

Polynomials and Transcendental Functions

Scalar arguments to transcendental functions such as exponential, trigonometric and logarithmic functions, or to inhomogeneous polynomials, must be dimensionless quantities. (Note: this requirement is somewhat relaxed in Siano's orientational analysis described below, in which the square of certain dimensioned quantities are dimensionless.)

While most mathematical identities about dimensionless numbers translate in a straightforward manner to dimensional quantities, care must be taken with logarithms of ratios: the identity log(a/b) = log a – log b, where the logarithm is taken in any base, holds for dimensionless numbers a and b, but it does not hold if a and b are dimensional, because in this case the left-hand side is well-defined but the right-hand side is not.

Similarly, while one can evaluate monomials (x^n) of dimensional quantities, one cannot evaluate polynomials of mixed degree with dimensionless coefficients on dimensional quantities: for x^2, the expression $(3\ m)^2 = 9\ m^2$ makes sense (as an area), while for $x^2 + x$, the expression $(3\ m)^2 + 3\ m = 9\ m^2 + 3$ m does not make sense.

However, polynomials of mixed degree can make sense if the coefficients are suitably chosen physical quantities that are not dimensionless.

For example,

$$\frac{1}{2}\cdot\left(-32\frac{\text{foot}}{\text{second}^2}\right)\cdot t^2+\left(500\frac{\text{foot}}{\text{second}}\right)\cdot t.$$

This is the height to which an object rises in time t if the acceleration of gravity is 32 feet per second per second and the initial upward speed is 500 feet per second. It is not even necessary for t to be in seconds. For example, suppose t = 0.01 minutes. Then the first term would be

$$\frac{1}{2}\cdot\left(-32\frac{\text{foot}}{\text{second}^2}\right)\cdot(0.01\ \text{minute})^2$$

$$=\frac{1}{2}\cdot-32\cdot(0.01^2)\left(\frac{\text{minute}}{\text{second}}\right)^2\cdot\text{foot}$$

$$=\frac{1}{2}\cdot-32\cdot(0.01^2)\cdot 60^2\cdot\text{foot}.$$

Incorporating Units

The value of a dimensional physical quantity Z is written as the product of a unit [Z] within the dimension and a dimensionless numerical factor, n.

$$Z=n\times[Z]=n[Z]$$

When like-dimensioned quantities are added or subtracted or compared, it is convenient to express them in consistent units so that the numerical values of these quantities may be directly added or subtracted. But, in concept, there is no problem adding quantities of the same dimension expressed in different units. For example, 1 meter added to 1 foot is a length, but one cannot derive that length by simply adding 1 and 1. A conversion factor, which is a ratio of like-dimensioned quantities and is equal to the dimensionless unity, is needed:

$$1\ \text{ft}=0.3048\ \text{m}\ \text{ is identical to }\ 1=\frac{0.3048\ \text{m}}{1\ \text{ft}}.$$

The factor $0.3048\frac{\text{m}}{\text{ft}}$ is identical to the dimensionless 1, so multiplying by this conversion factor changes nothing. Then when adding two quantities of like dimension, but expressed in different units, the appropriate conversion factor, which is essentially the dimensionless 1, is used to convert the quantities to identical units so that their numerical values can be added or subtracted.

Only in this manner is it meaningful to speak of adding like-dimensioned quantities of differing units.

Position vs Displacement

Some discussions of dimensional analysis implicitly describe all quantities as mathematical vectors. (In mathematics scalars are considered a special case of vectors; vectors can be added to or subtracted from other vectors, and, inter alia, multiplied or divided by scalars. If a vector is used to define a position, this assumes an implicit point of reference: an origin.

While this is useful and often perfectly adequate, allowing many important errors to be caught, it can fail to model certain aspects of physics. A more rigorous approach requires distinguishing between position and displacement (or moment in time versus duration, or absolute temperature versus temperature change).

Consider points on a line, each with a position with respect to a given origin, and distances among them. Positions and displacements all have units of length, but their meaning is not interchangeable:

- adding two displacements should yield a new displacement (walking ten paces then twenty paces gets you thirty paces forward),
- adding a displacement to a position should yield a new position (walking one block down the street from an intersection gets you to the next intersection),
- subtracting two positions should yield a displacement,
- but one may not add two positions.

This illustrates the subtle distinction between affine quantities (ones modelled by an affine space, such as position) and vector quantities (ones modelled by a vector space, such as displacement).

- Vector quantities may be added to each other, yielding a new vector quantity, and a vector quantity may be added to a suitable affine quantity (a vector space acts on an affine space), yielding a new affine quantity.
- Affine quantities cannot be added, but may be subtracted, yielding relative quantities which are vectors, and these relative differences may then be added to each other or to an affine quantity.

Properly then, positions have dimension of affine length, while displacements have dimension of vector length. To assign a number to an affine unit, one must not only choose a unit of measurement, but also a point of reference, while to assign a number to a vector unit only requires a unit of measurement.

Thus some physical quantities are better modelled by vectorial quantities while others tend to require affine representation, and the distinction is reflected in their dimensional analysis.

This distinction is particularly important in the case of temperature, for which the numeric value of absolute zero is not the origin 0 in some scales. For absolute zero,

$$0 \text{ K} = -273.15 \text{ °C} = -459.67 \text{ °F} = 0 \text{ °R},$$

but for temperature differences,

$$1 \text{ K} = 1 \text{ °C} \neq 1 \text{ °F} = 1 \text{ °R}.$$

(Here °R refers to the Rankine scale, not the Réaumur scale). Unit conversion for temperature differences is simply a matter of multiplying by, e.g., 1 °F / 1 K. But because some of these scales have origins that do not correspond to absolute zero, conversion from one temperature scale to another requires accounting for that. As a result, simple dimensional analysis can lead to errors if it is ambiguous whether 1 *K* means the absolute temperature equal to –272.15 °C, or the temperature difference equal to 1 °C.

Orientation and Frame of Reference

Similar to the issue of a point of reference is the issue of orientation: a displacement in 2 or 3 dimensions is not just a length, but is a length together with a direction. (This issue does not arise in 1 dimension, or rather is equivalent to the distinction between positive and negative.) Thus, to compare or combine two dimensional quantities in a multi-dimensional space, one also needs an orientation: they need to be compared to a frame of reference.

This leads to the extensions discussed below, namely Huntley's directed dimensions and Siano's orientational analysis.

A More Complex Example: Energy of a Vibrating Wire

Consider the case of a vibrating wire of length ℓ (L) vibrating with an amplitude A (L). The wire has a linear density ρ (M/L) and is under tension s (ML/T^2), and we want to know the energy E (ML^2/T^2) in the wire. Let π_1 and π_2 be two dimensionless products of powers of the variables chosen, given by

$$\pi_1 = E / As$$

$$\pi_2 = \ell / A.$$

The linear density of the wire is not involved. The two groups found can be combined into an equivalent form as an equation

$$F(E / As, \ell / A) = 0,$$

where F is some unknown function, or, equivalently as

$$E = Asf(\ell / A),$$

where f is some other unknown function. Here the unknown function implies that our solution is now incomplete, but dimensional analysis has given us something that may not have been obvious: the energy is proportional to the first power of the tension. Barring further analytical analysis, we might proceed to experiments to discover the form for the unknown function *f*. But our experiments are simpler than in the absence of dimensional analysis. We'd perform none to verify that the energy is proportional to the tension. Or perhaps we might guess that the energy is proportional to ℓ, and so infer that E = ℓ s. The power of dimensional analysis as an aid to experiment and forming hypotheses becomes evident.

The power of dimensional analysis really becomes apparent when it is applied to situations, unlike those given above, that are more complicated, the set of variables involved are not apparent, and the underlying equations hopelessly complex. Consider, for example, a small pebble sitting on the bed of a river. If the river flows fast enough, it will actually raise the pebble and cause it to flow along with the water. At what critical velocity will this occur? Sorting out the guessed variables is not so easy as before. But dimensional analysis can be a powerful aid in understanding problems like this, and is usually the very first tool to be applied to complex problems where the underlying equations and constraints are poorly understood. In such cases, the answer may depend on a dimensionless number such as the Reynolds number, which may be interpreted by dimensional analysis.

Extensions

Huntley's Extension: Directed Dimensions

Huntley (Huntley 1967) has pointed out that it is sometimes productive to refine our concept of dimension. Two possible refinements are:

- The magnitude of the components of a vector are to be considered dimensionally distinct. For example, rather than an undifferentiated length unit L, we may have L_x represent length in the x direction, and so forth. This requirement stems ultimately from the requirement that each component of a physically meaningful equation (scalar, vector, or tensor) must be dimensionally consistent.

- Mass as a measure of quantity is to be considered dimensionally distinct from mass as a measure of inertia.

As an example of the usefulness of the first refinement, suppose we wish to calculate the distance a cannonball travels when fired with a vertical velocity component V_y and a horizontal velocity component V_x, assuming it is fired on a flat surface. Assuming no use of directed lengths, the quantities of interest are then V_x, V_y, both dimensioned as L/T, R, the distance travelled, having dimension L, and g the downward acceleration of gravity, with dimension L/T^2

With these four quantities, we may conclude that the equation for the range R may be written:

$$R \propto V_x^a V_y^b g^c.$$

Or dimensionally

$$L = (L/T)^{a+b} (L/T^2)^c$$

from which we may deduce that $a+b+c=1$ and $a+b+2c=0$, which leaves one exponent undetermined. This is to be expected since we have two fundamental quantities L and T and four parameters, with one equation.

If, however, we use directed length dimensions, then V_x will be dimensioned as L_x/T, V_y as L_y/T, R as and L_x g as L_y/T^2. The dimensional equation becomes:

$$L_x = (L_x/T)^a (L_y/T)^b (L_y/T^2)^c$$

and we may solve completely as $a=1, b=1$ and $c=-1$. The increase in deductive power gained by the use of directed length dimensions is apparent.

In a similar manner, it is sometimes found useful (e.g., in fluid mechanics and thermodynamics) to distinguish between mass as a measure of inertia (inertial mass), and mass as a measure of quantity (substantial mass). For example, consider the derivation of Poiseuille's Law. We wish to find the rate of mass flow of a viscous fluid through a circular pipe. Without drawing distinctions between inertial and substantial mass we may choose as the relevant variables

- $\dot{m}$ the mass flow rate with dimensions M / T
- p_x the pressure gradient along the pipe with dimensions M / L^2T^2
- ρ the density with dimensions M / L^3
- η the dynamic fluid viscosity with dimensions M / LT
- r the radius of the pipe with dimensions L

There are three fundamental variables so the above five equations will yield two dimensionless variables which we may take to be $\pi_1 = \dot{m} / \eta r$ and $\pi_2 = p_x \rho r^5 / \dot{m}^2$ and we may express the dimensional equation as

$$C = \pi_1 \pi_2^a = \left(\frac{\dot{m}}{\eta r}\right)\left(\frac{p_x \rho r^5}{\dot{m}^2}\right)^a$$

where C and a are undetermined constants. If we draw a distinction between inertial mass with dimensions M_i and substantial mass with dimensions M_s, then mass flow rate and density will use substantial mass as the mass parameter, while the pressure gradient and coefficient of viscosity will use inertial mass. We now have four fundamental parameters, and one dimensionless constant, so that the dimensional equation may be written:

$$C = \frac{p_x \rho r^4}{\eta \dot{m}}$$

where now only C is an undetermined constant (found to be equal to $\pi / 8$ by methods outside of dimensional analysis). This equation may be solved for the mass flow rate to yield Poiseuille's law.

Siano's Extension: Orientational Analysis

Huntley's extension has some serious drawbacks:

- It does not deal well with vector equations involving the cross product,
- nor does it handle well the use of angles as physical variables.

It also is often quite difficult to assign the L, L_x, L_y, L_z, symbols to the physical variables involved in the problem of interest. He invokes a procedure that involves the "symmetry" of the physical problem. This is often very difficult to apply reliably: It is unclear as to what parts of the problem that the notion of "symmetry" is being invoked. Is it the symmetry of the physical body that forces are acting upon, or to the points, lines or areas at which forces are being applied? What if more than one body is involved with different symmetries? Consider the spherical bubble attached to a cylindrical tube, where one wants the flow rate of air as a function of the pressure difference in the two parts. What are the Huntley extended dimensions of the viscosity of the air contained in the connected parts? What are the extended dimensions of the pressure of the two parts? Are they the same or different? These difficulties are responsible for the limited application of Huntley's addition to real problems.

Angles are, by convention, considered to be dimensionless variables, and so the use of angles as physical variables in dimensional analysis can give less meaningful results. As an example, consider the projectile problem mentioned above. Suppose that, instead of the *x*- and *y*-components of the initial velocity, we had chosen the magnitude of the velocity v and the angle θ at which the projectile was fired. The angle is, by convention, considered to be dimensionless, and the magnitude of a vector has no directional quality, so that no dimensionless variable can be composed of the four variables *g, v, R,* and θ. Conventional analysis will correctly give the powers of *g* and *v*, but will give no information concerning the dimensionless angle θ.

Siano (1985-I, 1985-II) has suggested that the directed dimensions of Huntley be replaced by using orientational symbols $1_x\ 1_y\ 1_z$ to denote vector directions, and an orientationless symbol 1_0. Thus, Huntley's L_x becomes $L\ 1_x$ with L specifying the dimension of length, and 1_x specifying the orientation. Siano further shows that the orientational symbols have an algebra of their own. Along with the requirement that $1_i^{-1} = 1_i$, the following multiplication table for the orientation symbols results:

	$\mathbf{1_0}$	$\mathbf{1_x}$	$\mathbf{1_y}$	$\mathbf{1_z}$
$\mathbf{1_0}$	1_0	1_x	1_y	1_z
$\mathbf{1_x}$	1_x	1_0	1_z	1_y
$\mathbf{1_y}$	1_y	1_z	1_0	1_x
$\mathbf{1_z}$	1_z	1_y	1_x	1_0

Note that the orientational symbols form a group (the Klein four-group or "Viergruppe"). In this system, scalars always have the same orientation as the identity element, independent of the "symmetry of the problem." Physical quantities that are vectors have the orientation expected: a force or a velocity in the z-direction has the orientation of 1_z. For angles, consider an angle θ that lies in the z-plane. Form a right triangle in the *z* plane with θ being one of the acute angles. The side of the right triangle adjacent to the angle then has an orientation 1_x and the side opposite has an orientation 1_y. Then, since tan(θ) = $1_y/1_x$ = θ + ... we conclude that an angle in the *xy* plane must have an orientation $1_y/1_x = 1_z$, which is not unreasonable. Analogous reasoning forces the conclusion that sin(θ) has orientation 1_z while cos(θ) has orientation 1_0. These are different, so one concludes (correctly), for example, that there are no solutions of physical equations that are of the form a cos(θ)+*b* sin(θ) , where a and b are real scalars. Note that an expression such as $\sin(\theta+\pi/2)=\cos(\theta)$ is not dimensionally inconsistent since it is a special case of the sum of angles formula and should properly be written:

$$\sin(a1_z+b1_z)=\sin(a1_z)\cos(b1_z)+\sin(b1_z)\cos(a1_z)$$

which for $a=\theta$ and $b=\pi/2$ yields $\sin(\theta 1_z+(\pi/2)1_z)=1_z\cos(\theta 1_z)$. Physical quantities may be expressed as complex numbers (e.g. $e^{i\theta}$) which imply that the complex quantity i has an orientation equal to that of the angle it is associated with (1_z in the above example).

The assignment of orientational symbols to physical quantities and the requirement that physical equations be orientationally homogeneous can actually be used in a way that is similar to dimensional analysis to derive a little more information about acceptable solutions of physical problems. In this approach one sets up the dimensional equation and solves it as far as one can. If the lowest power of a physical variable is fractional, both sides of the solution is raised to a power such that all powers are integral. This puts it into "normal form". The orientational equation is then solved to give a more restrictive condition on the unknown powers of the orientational symbols, arriving at a solution that is more complete than the one that dimensional analysis alone gives. Often the added information is that one of the powers of a certain variable is even or odd.

As an example, for the projectile problem, using orientational symbols, θ, being in the xy-plane will thus have dimension 1_z and the range of the projectile R will be of the form:

$$R=g^a v^b \theta^c \text{ which means } L1_x \sim \left(\frac{L1_y}{T^2}\right)^a \left(\frac{L}{T}\right)^b 1_z^c.$$

Dimensional homogeneity will now correctly yield a = –1 and *b* = 2, and orientational homogeneity requires that *c* be an odd integer. In fact the required function of theta will be sin(θ)cos(θ) which is a series of odd powers of θ.

It is seen that the Taylor series of sin(θ) and cos(θ) are orientationally homogeneous using the above multiplication table, while expressions like cos(θ) + sin(θ) and exp(θ) are not, and are (correctly) deemed unphysical.

It should be clear that the multiplication rule used for the orientational symbols is not the same as that for the cross product of two vectors. The cross product of two identical vectors is zero, while the product of two identical orientational symbols is the identity element.

Dimensionless Concepts

Constants

The dimensionless constants that arise in the results obtained, such as the C in the Poiseuille's Law problem and the *K* in the spring problems discussed above come from a more detailed analysis of the underlying physics, and often arises from integrating some differential equation. Dimensional analysis itself has little to say about these constants, but it is useful to know that they very often have a magnitude of order unity. This observation can allow one to sometimes make "back of the envelope" calculations about the phenomenon of interest, and therefore be able to more efficiently design experiments to measure it, or to judge whether it is important, etc.

Formalisms

Paradoxically, dimensional analysis can be a useful tool even if all the parameters in the underlying theory are dimensionless, e.g., lattice models such as the Ising model can be used to study phase transitions and critical phenomena. Such models can be formulated in a purely dimensionless way. As we approach the critical point closer and closer, the distance over which the variables in the lattice model are correlated (the so-called correlation length, ξ) becomes larger and larger. Now, the correlation length is the relevant length scale related to critical phenomena, so one can, e.g., surmise on "dimensional grounds" that the non-analytical part of the free energy per lattice site should be $\sim 1/\xi^d$ where d is the dimension of the lattice.

It has been argued by some physicists, e.g., Michael Duff, that the laws of physics are inherently dimensionless. The fact that we have assigned incompatible dimensions to Length, Time and Mass is, according to this point of view, just a matter of convention, borne out of the fact that before the advent of modern physics, there was no way to relate mass, length, and time to each other. The three independent dimensionful constants: c, $\hbar$ and G, in the fundamental equations of physics must then be seen as mere conversion factors to convert Mass, Time and Length into each other.

Just as in the case of critical properties of lattice models, one can recover the results of dimensional analysis in the appropriate scaling limit; e.g., dimensional analysis in mechanics can be derived by reinserting the constants $\hbar$, c, and G (but we can now consider them to be dimensionless) and demanding that a nonsingular relation between quantities exists in the limit $c \to \infty$, $\hbar \to 0$ and $G \to 0$. In problems involving a gravitational field the latter limit should be taken such that the field stays finite.

Natural Units

If c = $\hbar$ = 1, where c = luminal speed and $\hbar$ = Planck's reduced constant, and a suitable fixed unit of energy is chosen, then all quantities of length L, mass M and time T can be expressed (dimensionally) as a power of energy E, because length, mass and time can be expressed using speed v, action S, and energy E:

$$M = E / v^2, \quad L = Sv / E, \quad t = S / E$$

though speed and action are dimensionless (v = c = 1 and S = $\hbar$ = 1) – so the only remaining quantity with dimension is energy. In terms of powers of dimensions:

$$E^n = M^p L^q t^r = E^{p-q-r}$$

This particularly useful in particle physics and high energy physics, in which case the energy unit is the electron volt (eV). Dimensional checks and estimates become very simple in this system.

However, if electric charges and currents are involved, another unit to be fixed is for electric charge, normally the electron charge e though other choices are possible.

Quantity	***p,q,r powers of energy***			***n power of energy***
	p	q	r	n
Action S	1	2	–1	0
Speed v	0	1	–1	0
Mass M	1	0	0	1
Length L	0	1	0	–1
Time t	0	0	1	–1
Momentum p	1	1	–1	1
Energy E	1	2	–2	1

Time Series and Quantile Regression

Quantile regression is a type of regression analysis used in statistics and econometrics. Whereas the method of least squares results in estimates that approximate the conditional mean of the response variable given certain values of the predictor variables, quantile regression aims at estimating either the conditional median or other quantiles of the response variable.

Advantages and Applications

Quantile regression is desired if conditional quantile functions are of interest. One advantage of quantile regression, relative to the ordinary least squares regression, is that the quantile regression estimates are more robust against outliers in the response measurements. However, the main attraction of quantile regression goes beyond that. In practice we often prefer using different measures of central tendency and statistical dispersion to obtain a more comprehensive analysis of the relationship between variables.

In ecology, quantile regression has been proposed and used as a way to discover more useful predictive relationships between variables in cases where there is no relationship or only a weak relationship between the means of such variables. The need for and success of quantile regression in ecology has been attributed to the complexity of interactions between different factors leading to data with unequal variation of one variable for different ranges of another variable. Another application of quantile regression is in the areas of growth charts, where percentile curves are commonly used to screen for abnormal growth.

Mathematics

The mathematical forms arising from quantile regression are distinct from those arising in the method of least squares. The method of least squares leads to a consideration of problems in an inner product space, involving projection onto subspaces, and thus the problem of minimizing the squared errors can be reduced to a problem in numerical linear algebra. Quantile regression does not have this structure, and instead leads to problems in linear programming that can be solved by the simplex method.

History

The idea of estimating a median regression slope, a major theorem about minimizing sum of the absolute deviances and a geometrical algorithm for constructing median regression was proposed in 1760 by Ruder Josip Boskovia, a Jesuit Catholic priest from Dubrovnik. Median regression computations for larger data sets are quite tedious compared to the least squares method, for which reason it has historically generated a lack of popularity among statisticians, until the widespread use of computers in the latter part of the 20th century.

Quantiles

Let Y be a real valued random variable with cumulative distribution function $F_Y(y) = P(Y \le y)$. The τ th quantile of Y is given by

$$Q_Y(\tau) = F_Y^{-1}(\tau) = \inf\{y : F_Y(y) \ge \tau\}$$

where $\tau \in [0,1]$.

Define the loss function as $\rho_\tau(y) = | y(\tau - \mathbb{I}_{(y<0)}) |$, where $\mathbb{I}$ is an indicator variable. A specific quantile can be found by minimizing the expected loss of $Y - u$ with respect to u:

$$\min_u E(\rho_\tau(Y-u)) = \min_u (\tau - 1)\int_{-\infty}^{u}(y-u)dF_Y(y) + \tau\int_{u}^{\infty}(y-u)dF_Y(y).$$

This can be shown by setting the derivative of the expected loss function to 0 and letting q_τ be the solution of

$$0 = (1-\tau)\int_{-\infty}^{q_\tau} dF_Y(y) - \tau\int_{q_\tau}^{\infty} dF_Y(y).$$

This equation reduces to

$$0 = F_Y(q_\tau) - \tau,$$

and then to

$$F_Y(q_\tau) = \tau.$$

Hence q_τ is t th quantile of the random variable Y.

Example

Let Y be a discrete random variable that takes values 1,2,..,9 with equal probabilities. The task $\tau = 0.5$ is to find the median of Y, and hence the value is chosen. The expected loss, $L(u)$, is

$$L(u) = \frac{(\tau-1)}{9}\sum_{y_i<u}(y_i - u) + \frac{\tau}{9}\sum_{y_i\ge u}(y_i - u) = \frac{0.5}{9}\left(-\sum_{y_i<u}(y_i-u) + \sum_{y_i\ge u}(y_i-u)\right).$$

Since $0.5/9$ is a constant, it can be taken out of the expected loss function (this is only true if $\tau = 0.5$). Then, at u=3,

$$L(3) \propto \sum_{i=1}^{2} -(i-3) + \sum_{i=3}^{9}(i-3) = [(2+1) + (0+1+2+\ldots+6)] = 24.$$

Suppose that u is increased by 1 unit. Then the expected loss will be changed by $(3) - (6) = -3$ on changing u to 4. If, u=5, the expected loss is

$$L(5) \propto \sum_{i=1}^{4} i + \sum_{i=0}^{4} i = 20,$$

and any change in u will increase the expected loss. Thus u=5 is the median. The Table below shows the expected loss (divided by $0.5/9$) for different values of u.

u	1	2	3	4	5	6	7	8	9
Expected loss	36	29	24	21	20	21	24	29	36

Intuition

Consider $\tau = 0.5$ and let q be an initial guess for q_τ. The expected loss evaluated at q is

$$-0.5\int_{-\infty}^{q}(y-q)dF_Y(y) + 0.5\int_{q}^{\infty}(y-q)dF_Y(y).$$

In order to minimize the expected loss, we move the value of q a little bit to see whether the expect loss will rise or fall. Suppose we increase q by 1 unit. Then the change of expected loss would be

$$\int_{-\infty}^{q} 1dF_Y(y) - \int_{q}^{\infty} 1dF_Y(y).$$

The first term of the equation is $F_Y(q)$ and second term of the equation is $1 - F_Y(q)$. Therefore the change of expected loss function is negative if and only if $F_Y(q) < 0.5$, that is if and only if q is smaller than the median. Similarly, if we reduce q by 1 unit, the change of expected loss function is negative if and only if q is larger than the median.

In order to minimize the expected loss function, we would increase (decrease) $L(q)$ if q is smaller (larger) than the median, until q reaches the median. The idea behind the minimization is to count the number of points (weighted with the density) that are larger or smaller than q and then move q to a point where q is larger than τ% of the points.

Sample Quantile

The sample quantile can be obtained by solving the following minimization problem

$$\hat{q}_\tau = \arg\min_{q \in R} \sum_{i=1}^{n} \rho_\tau(y_i - q),$$

$$= \arg\min_{q \in R} \left[(1-\tau)\sum_{y_i < q}(y_i - q) + \tau \sum_{y_i \geq q}(y_i - q) \right].$$

The intuition is the same as for the population quantile.

Conditional Quantile and Quantile Regression

Suppose the t th conditional quantile function is $Q_{Y|X}(\tau) = X\beta_\tau$ Given the distribution function of Y, β_τ an be obtained by solving

$$\beta_\tau = \arg\min_{\beta \in R^k} E(\rho_\tau(Y - X\beta)).$$

Solving the sample analog gives the estimator of β.

$$\widehat{\beta_\tau} = \arg\min_{\beta \in R^k} \sum_{i=1}^{n} (\rho_\tau(Y_i - X\beta)).$$

Computation

The minimization problem can be reformulated as a linear programming problem

$$\min_{\beta^+, \beta^-, u^+, u^- \in R^{2k} \times R_+^{2n}} \tau 1nu^+ + (1-\tau)1nu^- \mid X(\beta^+ - \beta^-) + u^+ - u^- = Y,$$

where

$$\beta_j^+ = \max(\beta_j, 0), \quad \beta_j^- = -\min(\beta_j, 0), \quad u_j^+ = \max(u_j, 0),$$

$$u_j^- = -\min(u_j, 0).$$

Simplex methods or interior point methods can be applied to solve the linear programming problem.

Asymptotic Properties

For $\tau \in (0,1)$, under some regularity conditions, $\hat{\beta}_\tau$ is asymptotically normal:

$$\sqrt{n}(\hat{\beta}_\tau - \beta_\tau) \xrightarrow{d} N(0, \tau(1-\tau)D^{-1}\Omega_x D^{-1}),$$

where

$$D = E(f_Y(X\beta)XX') \text{ and } \Omega_x = E(X'X).$$

Direct estimation of the asymptotic variance-covariance matrix is not always satisfactory. Inference for quantile regression parameters can be made with the regression rank-score tests or with the bootstrap methods.

Equivariance

Equivariance to Reparameterization of Design

Let A be any $p \times p$ nonsingular matrix and

$$\hat{\beta}(\tau; Y, XA) = A^{-1}\hat{\beta}(\tau; Y, X).$$

Invariance to Monotone Transformations

If h is a nondecreasing function on 'R, the following invariance property applies:

$$h(Q_{Y|X}(\tau)) \equiv Q_{h(Y)|X}(\tau).$$

Bayesian Methods for Quantile Regression

Because quantile regression does not normally assume a parametric likelihood for the conditional distributions of Y | X, the Bayesian methods work with a working likelihood. A convenient choice is the asymmetric Laplacian likelihood, because the mode of the resulting posterior under a flat prior is the usual quantile regression estimates. The posterior inference, however, must be interpreted with care. Yang and He (2012) showed that one can have asymptotically valid posterior inference if the working likelihood is chosen to be the empirical likelihood.

Censored Quantile Regression

If the response variable is subject to censoring, the conditional mean is not identifiable without additional distributional assumptions, but the conditional quantile is often identifiable.

Biostatistical Applications

Biostatistics (or biometry) is the application of statistics to a wide range of topics in biology. The science of biostatistics encompasses the design of biological experiments, especially in medicine, pharmacy, agriculture and fishery; the collection, summarization, and analysis of data from those experiments; and the interpretation of, and inference from, the results. A major branch of this is medical biostatistics, which is exclusively concerned with medicine and health.

Biostatistics and the History of Biological Thought

Biostatistical reasoning and modelling were of critical importance to the foundation theories of modern biology. In the early 1900s, after the rediscovery of Mendel's work, the gaps in understanding between genetics and evolutionary Darwinism led to vigorous debate among biometricians, such as Walter Weldon and Karl Pearson, and Mendelians, such as Charles Davenport, William Bateson and Wilhelm Johannsen. By the 1930s, statisticians and models built on statistical reasoning had helped to resolve these differences and to produce the neo-Darwinian modern evolutionary synthesis.

The leading figures in the establishment of this synthesis all relied on statistics and developed its use in biology.

- Sir Ronald A. Fisher developed several basic statistical methods in support of his work The Genetical Theory of Natural Selection
- Sewall G. Wright used statistics in the development of modern population genetics
- J. B. S Haldane's book, The Causes of Evolution, reestablished natural selection as the premier mechanism of evolution by explaining it in terms of the mathematical consequences of Mendelian genetics.

These individuals and the work of other biostatisticians, mathematical biologists, and statistically inclined geneticists helped bring together evolutionary biology and genetics into a consistent, coherent whole that could begin to be quantitatively modelled.

In parallel to this overall development, the pioneering work of D'Arcy Thompson in On Growth and Form also helped to add quantitative discipline to biological study.

Despite the fundamental importance and frequent necessity of statistical reasoning, there may nonetheless have been a tendency among biologists to distrust or deprecate results which are not qualitatively apparent. One anecdote describes Thomas Hunt Morgan banning the Friden calculator from his department at Caltech, saying "Well, I am like a guy who is prospecting for gold along the banks of the Sacramento River in 1849. With a little intelligence, I can reach down and pick up big nuggets of gold. And as long as I can do that, I'm not going to let any people in my department waste scarce resources in placer mining."

Scope and Training Programs

Almost all educational programmes in biostatistics are at postgraduate level. They are most often found in schools of public health, affiliated with schools of medicine, forestry, or agriculture, or as a focus of application in departments of statistics.

In the United States, where several universities have dedicated biostatistics departments, many other top-tier universities integrate biostatistics faculty into statistics or other departments, such as epidemiology. Thus, departments carrying the name "biostatistics" may exist under quite different structures. For instance, relatively new biostatistics departments have been founded with a focus on bioinformatics and computational biology, whereas older departments, typically affiliated with schools of public health, will have more traditional lines of research involving epidemiological studies and clinical trials as well as bioinformatics.

In larger universities where both a statistics and a biostatistics department exist, the degree of integration between the two departments may range from the bare minimum to very close collaboration. In general, the difference between a statistics program and a biostatistics program is twofold: (i) statistics departments will often host theoretical/methodological research which are less common in biostatistics programs and (ii) statistics departments have lines of research that may include biomedical applications but also other areas such as industry (quality control), business and economics and biological areas other than medicine.

Recent Developments in Modern Biostatistics

The advent of modern computer technology and relatively cheap computing resources have enabled computer-intensive biostatistical methods like bootstrapping and resampling methods. Furthermore new biomedical technologies like microarrays, next generation sequencers (for genomics) and mass spectrometry (for proteomics) generate enormous amounts of (redundant) data that can only be analyzed with biostatistical methods. For example, a microarray can measure all the genes of the human genome simultaneously, but only a fraction of them will be differentially expressed in diseased vs. non-diseased states. One might encounter the problem of multicolinearity:

Due to high intercorrelation between the predictors (in this case say genes), the information of one predictor might be contained in another one. It could be that only 5% of the predictors are responsible for 90% of the variability of the response. In such a case, one would apply the biostatistical technique of dimension reduction (for example via principal component analysis). Classical statistical techniques like linear or logistic regression and linear discriminant analysis do not work well for high dimensional data (i.e. when the number of observations n is smaller than the number of features or predictors p: $n < p$). As a matter of fact, one can get quite high R^2-values despite very low predictive power of the statistical model. These classical statistical techniques (esp. least squares linear regression) were developed for low dimensional data (i.e. where the number of observations n is much larger than the number of predictors p: $n >> p$). In cases of high dimensionality, one should always consider an independent validation test set and the corresponding residual sum of squares (RSS) and R^2 of the validation test set, not those of the training set.

In recent times, random forests have gained popularity. This technique, invented by the statistician Leo Breiman, generates a lot of decision trees randomly and uses them for classification (In classification the response is on a nominal or ordinal scale, as opposed to regression where the response is on a ratio scale). Decision trees have of course the advantage that you can draw them and interpret them (even with a very basic understanding of mathematics and statistics). Random Forrests have thus been used for clinical decision support systems.

Gene Set Enrichment Analysis (GSEA) is a new method for analyzing biological high throughput experiments. With this method, one does not consider the perturbation of single genes but of whole (functionally related) gene sets. These gene sets might be known biochemical pathways or otherwise functionally related genes. The advantage of this approach is that it is more robust: It is more likely that a single gene is found to be falsely perturbed than it is that a whole pathway is falsely perturbed. Furthermore, one can integrate the accumulated knowledge about biochemical pathways (like the JAK-STAT signaling pathway) using this approach.

Applications of Biostatistics

- Public health, including epidemiology, health services research, nutrition, environmental health and healthcare policy & management.
- Design and analysis of clinical trials in medicine
- Assessment of severity state of a patient with prognosis of outcome of a disease.
- Population genetics, and statistical genetics in order to link variation in genotype with a variation in phenotype. This has been used in agriculture to improve crops and farm animals (animal breeding). In biomedical research, this work can assist in finding candidates for gene alleles that can cause or influence predisposition to disease in human genetics
- Analysis of genomics data, for example from microarray or proteomics experiments. Often concerning diseases or disease stages.
- Ecology, ecological forecasting
- Biological sequence analysis
- Systems biology for gene network inference or pathways analysis.

Chapter 6

Nonstationary Time Series

In mathematics and statistics, a stationary process (or strict(ly) stationary process or strong(ly) stationary process) is a stochastic process whose joint probability distribution does not change when shifted in time. Consequently, parameters such as the mean and variance, if they are present, also do not change over time and do not follow any trends.

Stationarity is used as a tool in time series analysis, where the raw data is often transformed to become stationary; for example, economic data are often seasonal and/or dependent on a non-stationary price level. An important type of non-stationary process that does not include a trend-like behaviour is the cyclostationary process.

Note that a "stationary process" is not the same thing as a "process with a stationary distribution". Indeed there are further possibilities for confusion with the use of "stationary" in the context of stochastic processes; for example a "time-homogeneous" Markov chain is sometimes said to have "stationary transition probabilities". Besides, all stationary Markov random processes are time-homogeneous.

Non-stationary data, as a rule, are unpredictable and cannot be modelled or forecasted. The results obtained by using non-stationary time series may be spurious in that they may indicate a relationship between two variables where one does not exist. In order to receive consistent, reliable results, the non-stationary data needs to be transformed into stationary data. In contrast to the non-stationary process that has a variable variance and a mean that does not remain near, or returns to a long-run mean over time, the stationary process reverts around a constant long-term mean and has a constant variance independent of time.

Random Walk

A random walk is a mathematical formalization of a path that consists of a succession of random steps. For example, the path traced by a molecule as it travels in a liquid or a gas, the search path of a foraging animal, the price of a fluctuating stock and the financial status of a gambler can all be modelled as random walks, although they may not be truly random in reality. The term random walk was first introduced by Karl Pearson in 1905. Random walks have been used in many fields: ecology, economics, psychology, computer science, physics, chemistry, and biology. Random walks explain the observed behaviours of many processes in these fields, and thus serve as a fundamental model for the recorded stochastic activity.

Various different types of random walks are of interest. Often, random walks are assumed to be Markov chains or Markov processes, but other, more complicated walks are also of interest. Some random walks are on graphs, others on the line, in the plane, in higher dimensions, or even curved surfaces, while some random walks are on groups. Random walks also vary with regard to the time parameter. Often, the walk is in discrete time, and indexed by the natural numbers, as in $X_0, X_1, X_2, \ldots$. However, some walks take their steps at random times, and in that case the position X_t is defined for the continuum of times $t \geq 0$. Specific cases or limits of random walks include the Lévy flight. Random walks are related to the diffusion models and are a fundamental topic in discussions of Markov processes. Several properties of random walks, including dispersal distributions, first-passage times and encounter rates, have been extensively studied.

Lattice Random Walk

A popular random walk model is that of a random walk on a regular lattice, where at each step the location jumps to another site according to some probability distribution. In a simple random walk, the location can only jump to neighbouring sites of the lattice. In simple symmetric random walk on a locally finite lattice, the probabilities of the location jumping to each one of its immediate neighbours are the same. The best studied example is of random walk on the d-dimensional integer lattice (sometimes called the hypercubic lattice) $\mathbb{Z}^d$.

One-Dimensional Random Walk

An elementary example of a random walk is the random walk on the integer number line, $\mathbb{Z}$, which starts at 0 and at each step moves +1 or • 1 with equal probability.

This walk can be illustrated as follows. A marker is placed at zero on the number line and a fair coin is flipped. If it lands on heads, the marker is moved one unit to the right.

If it lands on tails, the marker is moved one unit to the left. After five flips, the marker could now be on 1, –1, 3, –3, 5, or –5. With five flips, three heads and two tails, in any order, will land on 1.

There are 10 ways of landing on 1 (by flipping three heads and two tails), 10 ways of landing on –1 (by flipping three tails and two heads), 5 ways of landing on 3 (by flipping four heads and one tail), 5 ways of landing on "3 (by flipping four tails and one head), 1 way of landing on 5 (by flipping five heads), and 1 way of landing on –5 (by flipping five tails).

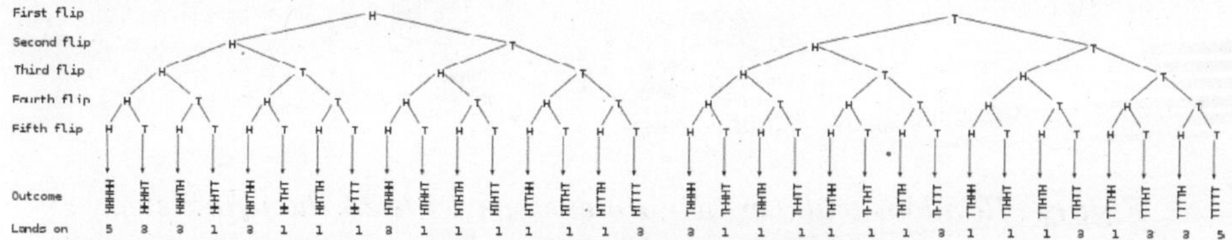

Figure: *All possible random walk outcomes after 5 flips of a fair coin*

Figure: *Random walk in two dimensions (animated version)*

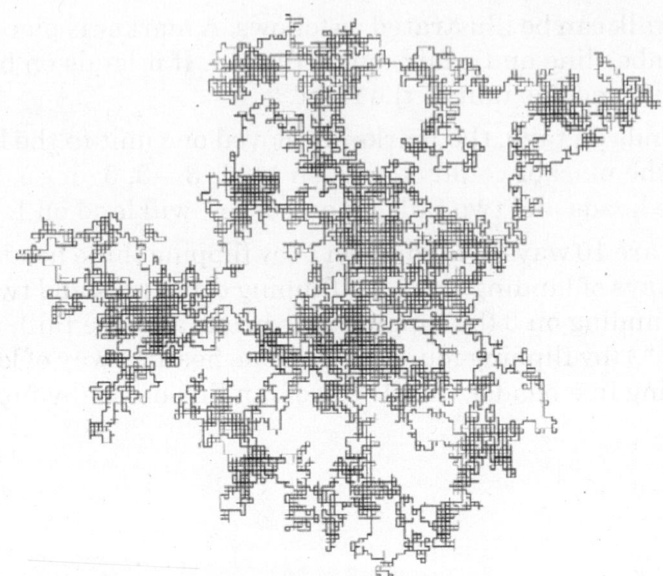

***Figure:** Random walk in two dimensions with 25 thousand steps (animated version)*

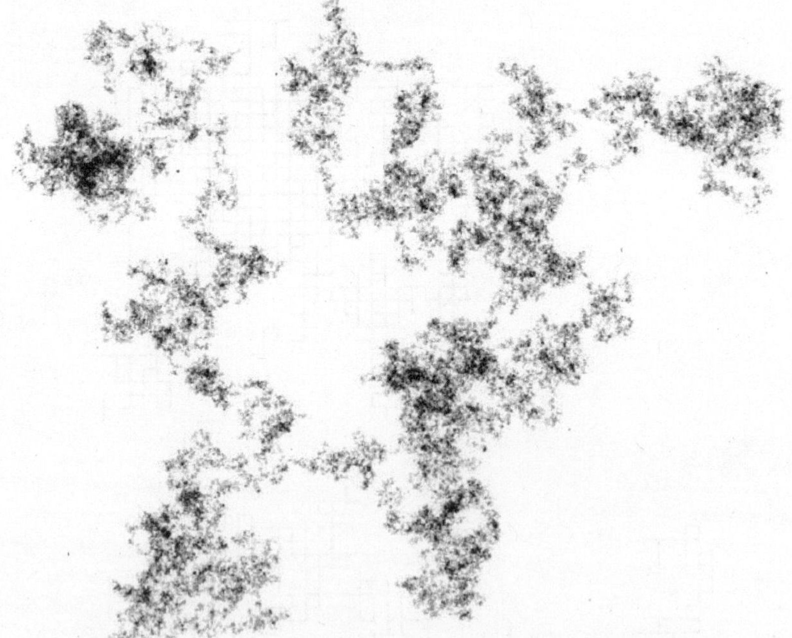

***Figure:** Random walk in two dimensions with two million even smaller steps. This image was generated in such a way that points that are more frequently traversed are darker. In the limit, for very small steps, one obtains Brownian motion.*

To define this walk formally, take independent random variables $Z_1, Z_2, \ldots$, where each variable is either 1 or –1, with a 50% probability for either value, and set $S_0 = 0$ and $S_n = \sum_{j=1}^{n} Z_j$. The series $\{S_n\}$ is called the simple random walk on . This series (the sum of the sequence of –1s and 1s) gives the distance walked, if each part of the walk is of length one. The expectation of is zero. That is, the mean of all coin flips approaches zero as the number of flips increases. This follows by the finite additivity property of expectation:

$$E(S_n) = \sum_{j=1}^{n} E(Z_j) = 0.$$

A similar calculation, using the independence of the random variables and the fact that $E(Z_n^2) = 1$, shows that:

$$E(S_n^2) = \sum_{j=1}^{n} E(Z_j^2) + \sum_{i=1}^{n} \sum_{j=1}^{n} 2E(Z_j Z_i) = n.$$

This hints that $E(|S_n|)$, the expected translation distance after n steps, should be of the order of $\sqrt{n}$. In fact,

$$\lim_{n \to \infty} \frac{E(|S_n|)}{\sqrt{n}} = \sqrt{\frac{2}{\pi}}.$$

This result shows that diffusion is ineffective for mixing because of the way the square root behaves for large N.

How many times will a random walk cross a boundary line if permitted to continue walking forever? A simple random walk on $\mathbb{Z}$ will cross every point an infinite number of times. This result has many names: the level-crossing phenomenon, recurrence or the gambler's ruin. The reason for the last name is as follows: a gambler with a finite amount of money will eventually lose when playing a fair game against a bank with an infinite amount of money. The gambler's money will perform a random walk, and it will reach zero at some point, and the game will be over.

If a and b are positive integers, then the expected number of steps until a one-dimensional simple random walk starting at 0 first hits b or $-a$ is ab. The probability that this walk will hit b before $-a$ is $a/(a+b)$, which can be derived from the fact that simple random walk is a martingale.

Some of the results mentioned above can be derived from properties of Pascal's triangle. The number of different walks of n steps where each step is +1 or –1 is 2^n. For the simple random walk, each of these walks are equally likely. In order for S_n to be equal to a number k it is necessary and sufficient that the number of +1 in the walk exceeds those of –1 by *k*. The number of walks which satisfy $S_n = k$ is equally the number of ways of choosing (*n* + *k*)/2 elements from an n element set, denoted $n\ (n+k)/2$. For this to be non-zero, it is necessary that *n* + *k* be an even number. Therefore, the probability that $S_n = k$ is equal to $2^{-n}\binom{n}{(n+k)/2}$. By representing entries of Pascal's triangle in terms of factorials and using Stirling's formula, one can obtain good estimates for these probabilities for large values of .

If the space is confined to + for brevity, the number of ways in which a random walk will land on any given number having five flips can be shown as {0,5,0,4,0,1}.

This relation with Pascal's triangle is demonstrated for small values of n. At zero turns, the only possibility will be to remain at zero. However, at one turn, there is one chance of landing on –1 or one chance of landing on 1. At two turns, a marker at 1 could move to 2 or back to zero. A marker at –1, could move to –2 or back to zero. Therefore, there is one chance of landing on –2, two chances of landing on zero, and one chance of landing on 2.

The central limit theorem and the law of the iterated logarithm describe important aspects of the behaviour of simple random walks on $\mathbb{Z}$. In particular, the former entails that as n increases, the probabilities (proportional to the numbers in each row) approach a normal distribution.

As a direct generalization, one can consider random walks on crystal lattices (infinite-fold abelian covering graphs over finite graphs). Actually it is possible to establish the central limit theorem and large deviation theorem in this setting.

As a Markov Chain

A one-dimensional random walk can also be looked at as a Markov chain whose state space is given by the integers $i = 0, \pm 1, \pm 2, \ldots$. For some number *p* satisfying $0 < p < 1$, the transition probabilities (the probability $P_{i,j}$ of moving from state *i* to state *j*) are given by

$$P_{i,i+1} = p = 1 - P_{i,i-1}.$$

Higher Dimensions

Imagine now a drunkard walking randomly in an idealized city. The city is effectively infinite and arranged in a square grid, and at every intersection, the drunkard chooses one of the four possible routes (including the one he came from) with equal probability. Formally, this is a random walk on the set of all points in the plane with integer coordinates.

Will the drunkard ever get back to his home from the bar? This is the 2-dimensional equivalent of the level crossing problem discussed above. It turns out that he almost surely will in a 2-dimensional random walk, but for 3 dimensions or higher, the probability of returning to the origin decreases as the number of dimensions increases. In 3 dimensions, the probability decreases to roughly 34%.

The trajectory of a random walk is the collection of sites it visited, considered as a set with disregard to when the walk arrived at the point. In one dimension, the trajectory is simply all points between the minimum height the walk achieved and the maximum (both are, on average, on the order of $\sqrt{n}$). In higher dimensions the set has interesting geometric properties. In fact, one gets a discrete fractal, that is a set which exhibits stochastic self-similarity on large scales, but on small scales one can observe "jaggedness" resulting from the grid on which the walk is performed. The two books of Lawler referenced below are a good source on this topic.

Relation to Wiener Process

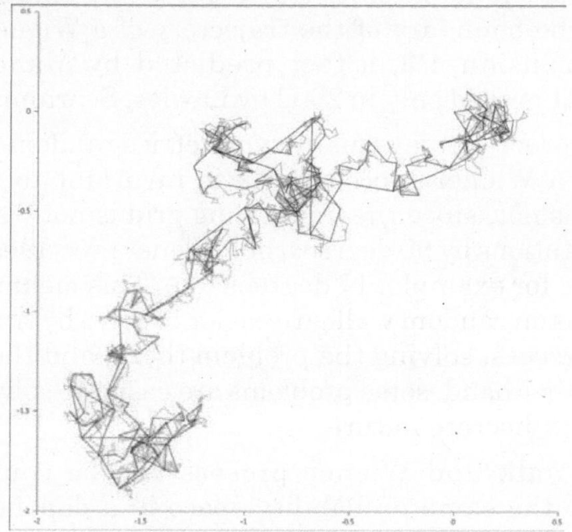

Figure: *Simulated steps approximating a Wiener process in two dimensions*

A Wiener process is a stochastic process with similar behaviour to Brownian motion, the physical phenomenon of a minute particle diffusing in a fluid. (Sometimes the Wiener process is called "Brownian motion", although this is strictly speaking a confusion of a model with the phenomenon being modelled.)

A Wiener process is the scaling limit of random walk in dimension 1. This means that if you take a random walk with very small steps you get an approximation to a Wiener process (and, less accurately, to Brownian motion). To be more precise, if the step size is ε, one needs to take a walk of length L/ε^2 to approximate a Wiener length of L. As the step size tends to 0 (and the number of steps increases proportionally) random walk converges to a Wiener process in an appropriate sense. Formally, if B is the space of all paths of length L with the maximum topology, and if M is the space of measure over B with the norm topology, then the convergence is in the space M. Similarly, a Wiener process in several dimensions is the scaling limit of random walk in the same number of dimensions.

A random walk is a discrete fractal (a function with integer dimensions; 1, 2, ...), but a Wiener process trajectory is a true fractal, and there is a connection between the two. For example, take a random walk until it hits a circle of radius r times the step length. The average number of steps it performs is r^2. This fact is the discrete version of the fact that a Wiener process walk is a fractal of Hausdorff dimension 2.

In two dimensions, the average number of points the same random walk has on the boundary of its trajectory is $r^{4/3}$. This corresponds to the fact that the boundary of the trajectory of a Wiener process is a fractal of dimension 4/3, a fact predicted by Mandelbrot using simulations but proved only in 2000 by Lawler, Schramm and Werner.

A Wiener process enjoys many symmetries random walk does not. For example, a Wiener process walk is invariant to rotations, but random walk is not, since the underlying grid is not (random walk is invariant to rotations by 90 degrees, but Wiener processes are invariant to rotations by, for example, 17 degrees too). This means that in many cases, problems on random walk are easier to solve by translating them to a Wiener process, solving the problem there, and then translating back. On the other hand, some problems are easier to solve with random walks due to its discrete nature.

Random walk and Wiener process can be coupled, namely manifested on the same probability space in a dependent way that forces them to be quite close. The simplest such coupling is the Skorokhod embedding, but other, more precise couplings exist as well.

The convergence of a random walk toward the Wiener process is controlled by the central limit theorem. For a particle in a known fixed position at $t = 0$, the theorem tells us that after a large number of independent steps in the random walk, the walker's position is distributed according to a normal distribution of total variance:

$$\sigma^2 = \frac{t}{\delta t}\varepsilon^2,$$

where t is the time elapsed since the start of the random walk, is the size of a step of the random walk, and δt is the time elapsed between two successive steps.

This corresponds to the Green function of the diffusion equation that controls the Wiener process, which demonstrates that, after a large number of steps, the random walk converges toward a Wiener process.

In 3*D*, the variance corresponding to the Green's function of the diffusion equation is:

$$\sigma^2 = 6Dt$$

By equalizing this quantity with the variance associated to the position of the random walker, one obtains the equivalent diffusion coefficient to be considered for the asymptotic Wiener process toward which the random walk converges after a large number of steps:

$$D = \frac{\varepsilon^2}{6\delta t} \text{ (valid only in 3D)}$$

Remark: the two expressions of the variance above correspond to the distribution associated to the vector $\vec{R}$ that links the two ends of the random walk, in 3*D*. The variance associated to each component R_x, R_y or R_z is only one third of this value (still in 3*D*).

Gaussian Random Walk

A random walk having a step size that varies according to a normal distribution is used as a model for real-world time series data such as financial markets. The Black–Scholes formula for modelling option prices, for example, uses a Gaussian random walk as an underlying assumption.

Here, the step size is the inverse cumulative normal distribution $\Phi^{-1}(z,\mu,\sigma)$ where $0 \le z \le 1$ is a uniformly distributed random number, and μ and σ are the mean and standard deviations of the normal distribution, respectively.

If μ is nonzero, the random walk will vary about a linear trend. If v_s is the starting value of the random walk, the expected value after n steps will be $v_s + n\mu$. For the special case where μ is equal to zero, after n steps, the translation distance's probability distribution is given by $N(0, n\sigma^2)$, where $N()$ is the notation for the normal distribution, n is the number of steps, and σ is from the inverse cumulative normal distribution as given above.

Proof: The Gaussian random walk can be thought of as the sum of a series of independent and identically distributed random variables, X_i from the inverse cumulative normal distribution with mean equal zero and σ of the original inverse cumulative normal distribution:

$$Z = \sum_{i=0}^{n} X_i$$

but we have the distribution for the sum of two independent normally distributed random variables, $Z = X + Y$, is given by

$$N(\mu_X + \mu_Y, \sigma^2{}_X + \sigma^2{}_Y).$$

In our case, $\mu_X = \mu_Y = 0$ and $\sigma^2{}_X = \sigma^2{}_Y = \sigma^2$ yield

$$N(0, 2\sigma^2)$$

By induction, for n steps we have

$$Z \sim N(0, n\sigma^2).$$

For steps distributed according to any distribution with zero mean and a finite variance (not necessarily just a normal distribution), the root mean square translation distance after n steps is

$$\sqrt{E\,|\,S_n^2\,|} = \sigma\sqrt{n}.$$

But for the Gaussian random walk, this is just the standard deviation of the translation distance's distribution after n steps. Hence, if μ is equal to zero, and since the root mean square(rms) translation distance is one standard deviation, there is 68.27% probability that the rms translation distance after n steps will fall between $\pm\,\sigma\sqrt{n}$. Likewise, there is 50% probability that the translation distance after n steps will fall between $\pm\,0.6745\sigma$.

Anomalous Diffusion

In disordered systems such as porous media and fractals may not be proportional to but to . The exponent is called the anomalous diffusion exponent and can be larger or smaller than 2. Anomalous diffusion may also be expressed as $\sigma_r{}^2 \sim Dt^\alpha$ where α is the anomaly parameter.

Number of Distinct Sites

The number of distinct sites visited by a single random walker has been studied extensively for square and cubic lattices and for fractals. This quantity is useful for the analysis of problems of trapping and kinetic reactions. It is also related to the vibrational density of states diffusion reactions processes and spread of populations in ecology.

The generalization of this problem to the number of distinct sites visited by random walkers, $S(t)$, has recently been studied for d-dimensional Euclidean lattices. The number of distinct sites visited by N walkers is not simply related to the number of distinct sites visited by each walker.

Applications

The following are some applications of random walk:

- In economics, the "random walk hypothesis" is used to model shares prices and other factors. Empirical studies found some deviations from this theoretical model, especially in short term and long term correlations.
- In population genetics, random walk describes the statistical properties of genetic drift
- In physics, random walks are used as simplified models of physical Brownian motion and diffusion such as the random movement of molecules in liquids and gases. Also in physics, random walks and some of the self interacting walks play a role in quantum field theory.
- In mathematical ecology, random walks are used to describe individual animal movements, to empirically support processes of biodiffusion, and occasionally to model population dynamics.
- In polymer physics, random walk describes an ideal chain. It is the simplest model to study polymers.
- In other fields of mathematics, random walk is used to calculate solutions to Laplace's equation, to estimate the harmonic measure, and for various constructions in analysis and combinatorics.
- In computer science, random walks are used to estimate the size of the Web. In the World Wide Web conference-2006, bar-yossef et al. published their findings and algorithms for the same.
- In image segmentation, random walks are used to determine the labels (i.e., "object" or "background") to associate with each

pixel. This algorithm is typically referred to as the random walker segmentation algorithm.

In all these cases, random walk is often substituted for Brownian motion.

- In brain research, random walks and reinforced random walks are used to model cascades of neuron firing in the brain.
- In vision science, fixational eye movements are well described by a random walk.
- In psychology, random walks explain accurately the relation between the time needed to make a decision and the probability that a certain decision will be made.
- Random walks can be used to sample from a state space which is unknown or very large, for example to pick a random page off the internet or, for research of working conditions, a random worker in a given country.
- When this last approach is used in computer science it is known as Markov Chain Monte Carlo or MCMC for short. Often, sampling from some complicated state space also allows one to get a probabilistic estimate of the space's size.

  The estimate of the permanent of a large matrix of zeros and ones was the first major problem tackled using this approach.
- Random walks have also been used to sample massive online graphs such as online social networks.
- In wireless networking, a random walk is used to model node movement.
- Motile bacteria engage in a biased random walk.
- Random walks are used to model gambling.
- In physics, random walks underlie the method of Fermi estimation.
- On the web, the Twitter website uses Random walks to make suggestions of who to follow

Variants of Random Walks

A number of types of stochastic processes have been considered that are similar to the pure random walks but where the simple structure is allowed to be more generalized.

The pure structure can be characterized by the steps being defined by independent and identically distributed random variables.

Random Walk on Graphs

A random walk of length k on a possibly infinite graph G with a root 0 is a stochastic process with random variables $X_1, X_2, \ldots, X_k$ such that $X_1 = 0$ and X_{i+1} is a vertex chosen uniformly at random from the neighbours of X_i. Then the number $p_{v,w,k}(G)$ is the probability that a random walk of length k starting at v ends at w. In particular, if G is a graph with root 0, $p_{0,0,2k}$ is the probability that a $2k$-step random walk returns to 0.

Assume now that our city is no longer a perfect square grid. When our drunkard reaches a certain junction he picks between the various available roads with equal probability. Thus, if the junction has seven exits the drunkard will go to each one with probability one seventh. This is a random walk on a graph. Will our drunkard reach his home? It turns out that under rather mild conditions, the answer is still yes. For example, if the lengths of all the blocks are between a and b (where a and b are any two finite positive numbers), then the drunkard will, almost surely, reach his home. Notice that we do not assume that the graph is planar, i.e. the city may contain tunnels and bridges. One way to prove this result is using the connection to electrical networks. Take a map of the city and place a one ohm resistor on every block. Now measure the "resistance between a point and infinity". In other words, choose some number R and take all the points in the electrical network with distance bigger than R from our point and wire them together. This is now a finite electrical network and we may measure the resistance from our point to the wired points. Take R to infinity. The limit is called the resistance between a point and infinity. It turns out that the following is true (an elementary proof can be found in the book by Doyle and Snell):

Theorem: a graph is transient if and only if the resistance between a point and infinity is finite. It is not important which point is chosen if the graph is connected.

In other words, in a transient system, one only needs to overcome a finite resistance to get to infinity from any point. In a recurrent system, the resistance from any point to infinity is infinite.

This characterization of recurrence and transience is very useful, and specifically it allows us to analyze the case of a city drawn in the plane with the distances bounded.

A random walk on a graph is a very special case of a Markov chain. Unlike a general Markov chain, random walk on a graph enjoys a property called time symmetry or reversibility. Roughly speaking, this property, also called the principle of detailed balance, means that the probabilities to traverse a given path in one direction or in the other have a very simple connection between them (if the graph is regular, they are just equal). This property has important consequences.

Starting in the 1980s, much research has gone into connecting properties of the graph to random walks. In addition to the electrical network connection described above, there are important connections to isoperimetric inequalities, functional inequalities such as Sobolev and Poincaré inequalities and properties of solutions of Laplace's equation. A significant portion of this research was focused on Cayley graphs of finitely generated groups. For example, the proof of Dave Bayer and Persi Diaconis that 7 riffle shuffles are enough to mix a pack of cards is in effect a result about random walk on the group Sn, and the proof uses the group structure in an essential way. In many cases these discrete results carry over to, or are derived from manifolds and Lie groups.

A good reference for random walk on graphs is the online book by Aldous and Fill. If the transition kernel $p(x, y)$ is itself random (based on an environment ω) then the random walk is called a "random walk in random environment". When the law of the random walk includes the randomness of ω, the law is called the annealed law; on the other hand, if is seen as fixed, the law is called a quenched law.

We can think about choosing every possible edge with the same probability as maximizing uncertainty (entropy) locally. We could also do it globally – in maximal entropy random walk (MERW) we want all paths to be equally probable, or in other words: for each two vertexes, each path of given length is equally probable. This random walk has much stronger localization properties.

Self-Interacting Random Walks

There are a number of interesting models of random paths in which each step depends on the past in a complicated manner. All are more complex for solving analytically than the usual random walk; still, the behaviour of any model of a random walker is obtainable using computers. Examples include:

- The self-avoiding walk (Madras and Slade 1996).

The self-avoiding walk of length *n* on *Z^d* is the random n-step path which starts at the origin, makes transitions only between adjacent sites in *Z^d*, never revisits a site, and is chosen uniformly among all such paths. In two dimensions, due to self-trapping, a typical self-avoiding walk is very short, while in higher dimension it grows beyond all bounds. This model has often been used in polymer physics (since the 1960s).

- The loop-erased random walk (Gregory Lawler).
- The reinforced random walk (Robin Pemantle 2007).
- The exploration process.
- The multiagent random walk.

Long-Range Correlated Walks

Long-range correlated time series are found in many biological, climatological and economic systems.

- Heartbeat records
- Non-coding DNA sequences
- Volatility time series of stocks
- Temperature records around the globe

Nonstationary Models for Time Series

The models presented so far are based on the stationarity assumption, that is, the mean and the variance of the underlying process are constant and the autocovariances depend only on the time lag. But many economic and business time series are nonstationary. Nonstationary time series can occur in many different ways. In particular, economic time series usually show time-changing levels, μ_t, and/or variances.

Null Hypothesis

In statistical inference of observed data of a scientific experiment, the null hypothesis refers to a general statement or default position that there is no relationship between two measured phenomena. Rejecting or disproving the null hypothesis – and thus concluding that there are grounds for believing that there is a relationship between two phenomena or that a potential treatment has a measurable effect – is a central task in the modern practice of science, and gives a precise sense in which a claim is capable of being proven false.

In statistical significance, the null hypothesis is often denoted H_0 (read "H-nought") and is generally assumed true until evidence indicates otherwise. The concept of a null hypothesis is used differently in two approaches to statistical inference. In the significance testing approach of Ronald Fisher, a null hypothesis is potentially rejected or disproved on the basis of data that is significant under its assumption, but never accepted or proved. In the hypothesis testing approach of Jerzy Neyman and Egon Pearson, a null hypothesis is contrasted with an alternative hypothesis, and these are distinguished on the basis of data, with certain error rates. Proponents of these two approaches criticize each other, though today a hybrid approach is widely practiced and presented in textbooks. This hybrid is in turn criticized as incorrect and incoherent. Statistical significance plays a pivotal role in statistical hypothesis testing where it is used to determine if a null hypothesis can be rejected or retained.

Principle

Hypothesis testing works by collecting data and measuring how likely the particular set of data is, assuming the null hypothesis is true, when the study is on a random representative sample. The null hypothesis assumes no relationship between variables in the population from which the sample is selected. If the data-set of a random representative sample is very unlikely relative to the null hypothesis, defined as being part of a class of sets of data that only rarely will be observed, the experimenter rejects the null hypothesis concluding it (probably) is false. This class of data-sets is usually specified via a test statistic which is designed to measure the extent of apparent departure from the null hypothesis.

The procedure works by assessing whether the observed departure measured by the test statistic is larger than a value defined so that the probability of occurrence of a more extreme value is small under the null hypothesis (usually in less than either 5% or 1% of similar data-sets in which the null hypothesis does hold). If the data do not contradict the null hypothesis, then only a weak conclusion can be made; namely that the observed data set provides no strong evidence against the null hypothesis. As the null hypothesis could be true or false, in this case, in some contexts this is interpreted as meaning that the data give insufficient evidence to make any conclusion, on others it means that there is no evidence to support changing from a currently useful regime to a different one.

For instance, a certain drug may reduce the chance of having a heart attack. Possible null hypotheses are “this drug does not reduce the chances of having a heart attack” or “this drug has no effect on the chances of having a heart attack”. The test of the hypothesis consists of administering the drug to half of the people in a study group as a controlled experiment. If the data show a statistically significant change in the people receiving the drug, the null hypothesis is rejected.

Testing for Differences

In scientific and medical research, null hypotheses play a major role in testing the significance of differences in treatment and control groups. This use, while widespread, offers several grounds for criticism, including straw man, Bayesian criticism and publication bias.

The typical null hypothesis at the outset of the experiment is that no difference exists between the control and experimental groups (for the variable being compared). Other possibilities include:

- that values in samples from a given population can be modelled using a certain family of statistical distributions.
- that the variability of data in different groups is the same, although they may be centred around different values.

Example

Given the test scores of two random samples of men and women, does one group differ from the other? A possible null hypothesis is that the mean male score is the same as the mean female score:

$$H_0: \mu_1 = \mu_2$$

where:

H_0 = the null hypothesis

μ_1 = the mean of population 1, and

μ_2 = the mean of population 2.

A stronger null hypothesis is that the two samples are drawn from the same population, such that the variance and shape of the distributions are also equal.

Terminology

Simple Hypothesis: Any hypothesis which specifies the population distribution completely. For such a hypothesis the sampling distribution of any statistic is a function of the sample size alone.

Composite Hypothesis: Any hypothesis which does not specify the population distribution completely. Example: A hypothesis specifying a normal distribution with a specified mean and an unspecified variance.

The simple/composite distinction was made by Neyman and Pearson.

Exact Hypothesis: Any hypothesis that specifies an exact parameter value. Example: $\mu = 100$. Synonym: point hypothesis.

Inexact Hypothesis: Those specifying a parameter range or interval. Examples: $\mu \leq 100$; $95 \leq \mu \leq 105$.

Fisher required an exact null hypothesis for testing.

A one-tailed hypothesis (AKA one-sided test) is an inexact hypothesis in which the value of a parameter is specified as being either:

- above or equal to a certain value, or
- below or equal to a certain value.

A One-Tailed Hypothesis is said to have Directionality.

Fisher's original (Lady tasting tea) example was a one-tailed test. The null hypothesis was symmetric. The odds of guessing all cups correctly was the same as guessing all cups incorrectly, but Fisher noted that only guessing correctly was compatible with the Lady's claim.

Goals of Null Hypothesis Tests

Statistical tests can be significance tests or hypothesis tests. There are many types of significance tests for one, two or more samples, for means, variances and proportions, paired or unpaired data, for different distributions, for large and small samples... All have null hypotheses. There are also at least 4 goals of null hypotheses for significance tests:

- Technical null hypotheses are used to verify statistical assumptions. Example: The residuals between the data and a statistical model cannot be distinguished from random noise. If true, there is no justification for complicating the model.
- Scientific null assumptions are used to directly advance a theory. Example: The angular momentum of the universe is zero. If not true, the theory of the early universe may need revision.
- Null hypotheses of homogeneity are used to verify that multiple experiments are producing consistent results. Example: The effect of a medication on the elderly is consistent with that of the general adult population. If true, this strengthens the general effectiveness conclusion and simplifies recommendations for use.
- Null hypotheses that assert the equality of effect of two or more alternative treatments, for example, a drug and a placebo, are used to reduce scientific claims based on statistical noise. This is the most popular null hypothesis; It is so popular that many statements about significant testing assume such null hypotheses.

Rejection of the null hypothesis is not necessarily the real goal of a significance tester. An adequate statistical model may be associated with a failure to reject the null; The model is adjusted until the null is not rejected. The numerous uses of significance testing were well known to Fisher who discussed many in his book written a decade before defining the null hypothesis.

A statistical significance test shares much mathematics with a confidence interval. They are mutually illuminating. A result is often significant when there is confidence in the sign of a relationship (the interval does not include 0). Whenever the sign of a relationship is important, statistical significance is a worthy goal. This also reveals weaknesses of significance testing: A result can be significant without a good estimate of the strength of a relationship; Significance can be a modest goal. A weak relationship can also achieve significance with enough data. Reporting both significance and confidence intervals is commonly recommended.

The varied uses of significance tests reduce the number of generalizations that can be made about all applications.

Choice of the Null Hypothesis

The choice of the null hypothesis is associated with sparse and inconsistent advice. Fisher mentioned few constraints on the choice and stated that many null hypotheses should be considered and that many tests are possible for each. The variety of applications and the diversity of goals suggests that the choice can be complicated. In many applications the formulation of the test is traditional. A familiarity with the range of tests available may suggest a particular null hypothesis and test. Formulating the null hypothesis is not automated; The calculations of significance testing usually are.

Caution: A statistical significance test is intended to test a hypothesis. If the hypothesis summarizes a set of data, there is no value in testing the hypothesis on that set of data. Example: If a study of last year's weather reports indicates that rain in a region falls primarily on weekends, it is only valid to test that null hypothesis on weather reports from any other year. Testing hypotheses suggested by the data is circular reasoning that proves nothing; It is a special limitation on the choice of the null hypothesis.

Routine advice: Start from the scientific hypothesis. Translate this to a statistical alternative hypothesis and proceed: "Because H_a expresses the effect that we wish to find evidence for, we often begin with H_a and then set up H_0 as the statement that the hoped-for effect is not present." This advice is reversed for modelling applications where we hope not to find evidence against the null.

A complex case example: The gold standard in clinical research is the randomised placebo controlled double blind clinical trial. But testing a new drug against a (medically ineffective) placebo may be unethical for a serious illness. Testing a new drug against an older medically effective drug raises fundamental philosophical issues regarding the goal of the test and the motivation of the experimenters. The standard "no difference" null hypothesis may reward the pharmaceutical company for gathering inadequate data. "Difference" is a better null hypothesis in this case, but statistical significance is not an adequate criterion for reaching a nuanced conclusion which requires a good numeric estimate of the drug's effectiveness. A "minor" or "simple" proposed change in the null hypothesis ((new vs old) rather than (new vs placebo)) can have a dramatic effect on the utility of a test for complex non-statistical reasons.

Directionality

The choice of null hypothesis (H_0) and consideration of directionality is critical. Consider the question of whether a tossed coin is fair (i.e. that on average it lands heads up 50% of the time). A potential null hypothesis is "this coin is not biased toward heads" (one-tail test). The experiment is to repeatedly toss the coin. A possible result of 5 tosses is 5 heads. Under this null hypothesis, the data are considered unlikely (with a fair coin, the probability of this is $1/2^5=3.1\%$ and the result would be even more unlikely if the coin were biased in favour of tails). The data refute the null hypothesis (that the coin is either fair or biased toward tails) and the conclusion is that the coin is biased towards heads.

Alternatively, the null hypothesis, "this coin is fair" could be examined by looking out for either too many tails or too many heads, and thus the types of outcomes that would tend to contradict this null hypothesis are those where a large number of heads or a large number of tails are observed. Thus a possible diagnostic outcome would be that all tosses yield the same outcome, and the probability of 5 of a kind is 6% under the null hypothesis. This is not statistically significant, preserving the null hypothesis in this case.

This example illustrates that the conclusion reached from a statistical test may depend on the precise formulation of the null and alternative hypotheses.

Fisher said, "the null hypothesis must be exact, that is free of vagueness and ambiguity, because it must supply the basis of the 'problem of distribution,' of which the test of significance is the solution", implying a more restrictive domain for H_0. According to this view, the null hypothesis must be numerically exact—it must state that a particular quantity or difference is equal to a particular number. In classical science, it is most typically the statement that there is no effect of a particular treatment; in observations, it is typically that there is no difference between the value of a particular measured variable and that of a prediction. The majority of null hypotheses in practice do not meet this "exactness" criterion. For example, consider the usual test that two means are equal where the true values of the variances are unknown—exact values of the variances are not specified.

Most statisticians believe that it is valid to state direction as a part of null hypothesis, or as part of a null hypothesis/alternative hypothesis pair. However, the results are not a full description of all the results of an experiment, merely a single result tailored to one particular purpose. For example, consider an H_0 that claims the population mean for a new treatment is an improvement on a well-established treatment with population mean = 10 (known from long experience), with the one-tailed alternative being that the new treatment's mean > 10. If the sample evidence obtained through x-bar equals –200 and the corresponding t-test statistic equals –50, the conclusion from the test would be that there is no evidence that the new treatmnent is better than the existing one: it would not report that it is markedly worse, but that is not what this particular test is looking for. To overcome any possible ambiguity in reporting the result of the test of a null hypothesis, it is best to indicate whether the test was two-sided and, if one-sided, to include the direction of the effect being tested.

The statistical theory required to deal with the simple cases of directionality dealt with here, and more complicated ones, makes use of the concept of an unbiased test.

The directionality of hypotheses is not always obvious. The explicit null hypothesis of Fisher's Lady tasting tea example was that the Lady had no such ability, which led to a symmetric probability distribution. The one-tailed nature of the test resulted from the one-tailed alternate hypothesis (a term not used by Fisher). The null hypothesis became implicitly one-tailed. The logical negation of the Lady's one-tailed claim was also one-tailed. (Claim: Ability > 0; Stated null: Ability = 0; Implicit null: Ability ≤ 0).

Pure arguments over the use of one-tailed tests are complicated by the variety of tests. Some tests (for instance the χ^2 goodness of fit test) are inherently one-tailed. Some probability distributions are asymmetric. The traditional tests of 3 or more groups are two-tailed.

Advice concerning the use of one-tailed hypotheses has been inconsistent and accepted practice varies among fields. The greatest objection to one-tailed hypotheses is their potential subjectivity. A non-significant result can sometimes be converted to a significant result by the use of a one-tailed hypothesis (as the fair coin test, at the whim of the analyst). The flip side of the argument: One-sided tests are less likely to ignore a real effect. One-tailed tests can suppress the publication of data that differs in sign from predictions. Objectivity was a goal of the developers of statistical tests.

Routine advice: Use one-tailed hypotheses by default: "If you do not have a specific direction firmly in mind in advance, use a two-sided alternative. Moreover, some users of statistics argue that we should always work with the two-sided alternative."

One alternative to this advice is to use three-outcome tests. It eliminates the issues surrounding directionality of hypotheses by testing twice, once in each direction and combining the results to produce three possible outcomes. Variations on this approach have a history, being suggested perhaps 10 times since 1950.

Disagreements over one-tailed tests flow from the philosophy of science. While Fisher was willing to ignore the unlikely case of the Lady guessing all cups of tea incorrectly (which may have been appropriate for the circumstances), medicine believes that a proposed treatment that kills patients is significant in every sense and should be reported and perhaps explained. Poor statistical reporting practices have contributed to disagreements over one-tailed tests. Statistical significance resulting from two-tailed tests is insensitive to the sign of the relationship; Reporting significance alone is inadequate.

"The treatment has an effect" is the uninformative result of a two-tailed test. "The treatment has a beneficial effect" is the more informative result of a one-tailed test. "The treatment has an effect, reducing the average length of hospitalization by 1.5 days" is the most informative report, combining a two-tailed significance test result with a numeric estimate of the relationship between treatment and effect. Explicitly reporting a numeric result eliminates a philosophical advantage of a one-tailed test. An underlying issue is the appropriate form of an experimental science without numeric predictive theories: A model of numeric results is more informative than a model of effect signs (positive, negative or unknown) which is more informative than a model of simple significance (non-zero or unknown); in the absence of numeric theory signs may suffice.

SpatioTemporal Time Series

A spatiotemporal database is a database that manages both space and time information. Common examples include:

- Tracking of moving objects, which typically can occupy only a single position at a given time.
- A database of wireless communication networks, which may exist only for a short timespan within a geographic region.
- An index of species in a given geographic region, where over time additional species may be introduced or existing species migrate or die out.
- Historical tracking of plate tectonic activity.

At first glance, spatiotemporal databases are an extension of spatial databases. A spatiotemporal database embodies spatial, temporal, and spatiotemporal database concepts, and captures spatial and temporal aspects of data and deals with

- geometry changing over time and/or
- location of objects moving over invariant geometry (known variously as moving objects databases or real-time locating systems)

However, although there exist numerous relational databases with spatial extensions, the spatiotemporal databases are not based on the relational model for practical reasons, chiefly among them that the data is multi-dimensional and capturing complex structures and behaviours. As of 2008, there are no RDBMS products with spatiotemporal extensions. There are some products such as the open-source TerraLib which use a middleware approach storing their data in a relational database. Unlike in the pure spatial domain, there are however no official or de facto standards for spatio-temporal data models and their querying. In general, the theory of this area is also less well-developed. Another approach is the constraint database system such as MPLQ (Management of Linear Programming Queries).

Continuous Time Series

In probability theory and statistics, a continuous-time stochastic process, or a continuous-space-time stochastic process is a stochastic process for which the index variable takes a continuous set of values, as contrasted with a discrete-time process for which the index variable takes only distinct values. An alternative terminology uses continuous parameter as being more inclusive.

A more restricted class of processes are the continuous stochastic processes: here the term often (but not always implies both that the index variable is continuous and that sample paths of the process are continuous. Given the possible confusion, caution is needed.

Continuous-time stochastic processes that are constructed from discrete-time processes via a waiting time distribution are called continuous-time random walks.

Spectral and Wavelet Methods

Spectral methods are a class of techniques used in applied mathematics and scientific computing to numerically solve certain differential equations, often involving the use of the Fast Fourier Transform. The idea is to write the solution of the differential equation as a sum of certain "basis functions" (for example, as a Fourier series which is a sum of sinusoids) and then to choose the coefficients in the sum in order to satisfy the differential equation as well as possible.

Spectral methods and finite element methods are closely related and built on the same ideas; the main difference between them is that spectral methods use basis functions that are nonzero over the whole domain, while finite element methods use basis functions that are nonzero only on small subdomains. In other words, spectral methods take on a global approach while finite element methods use a local approach. Partially for this reason, spectral methods have excellent error properties, with the so-called "exponential convergence" being the fastest possible, when the solution is smooth. However, there are no known three-dimensional single domain spectral shock capturing results. In the finite element community, a method where the degree of the elements is very high or increases as the grid parameter h decreases to zero is sometimes called a spectral element method.

Spectral methods can be used to solve ordinary differential equations (ODEs), partial differential equations (PDEs) and eigenvalue problems involving differential equations. When applying spectral methods to time-dependent PDEs, the solution is typically written as a sum of basis functions with time-dependent coefficients; substituting this in the PDE yields a system of ODEs in the coefficients which can be solved using any numerical method for ODEs. Eigenvalue problems for ODEs are similarly converted to matrix eigenvalue problems

Spectral methods were developed in a long series of papers by Steven Orszag starting in 1969 including, but not limited to, Fourier series methods for periodic geometry problems, polynomial spectral methods for finite and unbounded geometry problems, pseudospectral methods for highly nonlinear problems, and spectral iteration methods for fast solution of steady state problems. The implementation of the spectral method is normally accomplished either with collocation or a Galerkin or a Tau approach.

Spectral methods are computationally less expensive than finite element methods, but become less accurate for problems with complex geometries and discontinuous coefficients. This increase in error is a consequence of the Gibbs phenomenon.

A Relationship with the Spectral Element Method

One can show that if g is infinitely differentiable, then the numerical algorithm using Fast Fourier Transforms will converge faster than any polynomial in the grid size h. That is, for any $n>0$, there is a $C < \infty$ such that the error is less than Ch^n for all sufficiently small values of h. We say that the spectral method is of order n, for every n>0.

Because a spectral element method is a finite element method of very high order, there is a similarity in the convergence properties. However, whereas the spectral method is based on the eigendecomposition of the particular boundary value problem, the spectral element method does not use that information and works for arbitrary elliptic boundary value problems.

Wavelet Theory

Wavelet theory is applicable to several subjects. All wavelet transforms may be considered forms of time-frequency representation for continuous-time (analog) signals and so are related to harmonic analysis. Almost all practically useful discrete wavelet transforms use discrete-time filterbanks. These filter banks are called the wavelet and scaling coefficients in wavelets nomenclature. These filterbanks may contain either finite impulse response (FIR) or infinite impulse response (IIR) filters. The wavelets forming a continuous wavelet transform (CWT) are subject to the uncertainty principle of Fourier analysis respective sampling theory: Given a signal with some event in it, one cannot assign simultaneously an exact time and frequency response scale to that event. The product of the uncertainties of time and frequency response scale has a lower bound. Thus, in the scaleogram of a continuous wavelet transform of this signal, such an event marks an entire region in the time-scale plane, instead of just one point. Also, discrete wavelet bases may be considered in the context of other forms of the uncertainty principle.

Wavelet transforms are broadly divided into three classes: continuous, discrete and multiresolution-based.

Mother Wavelet

For practical applications, and for efficiency reasons, one prefers continuously differentiable functions with compact support as mother (prototype) wavelet (functions). However, to satisfy analytical requirements (in the continuous WT) and in general for theoretical reasons, one chooses the wavelet functions from a subspace of the space $L^1(\mathbb{R}) \cap L^2(\mathbb{R})$. This is the space of measurable functions that are absolutely and square integrable:

$$\int_{-\infty}^{\infty} |\psi(t)|\,dt < \infty \text{ and } \int_{-\infty}^{\infty} |\psi(t)|^2\,dt < \infty.$$

Being in this space ensures that one can formulate the conditions of zero mean and square norm one:

$$\int_{-\infty}^{\infty} \psi(t)\,dt = 0 \text{ is the condition for zero mean, and}$$

$$\int_{-\infty}^{\infty} |\psi(t)|^2\,dt = 1 \text{ is the condition for square norm one.}$$

For ψ to be a wavelet for the continuous wavelet transform, the mother wavelet must satisfy an admissibility criterion (loosely speaking, a kind of half-differentiability) in order to get a stably invertible transform.

For the discrete wavelet transform, one needs at least the condition that the wavelet series is a representation of the identity in the space $L^2(R)$. Most constructions of discrete WT make use of the multiresolution analysis, which defines the wavelet by a scaling function. This scaling function itself is solution to a functional equation.

In most situations it is useful to restrict ψ to be a continuous function with a higher number M of vanishing moments, i.e. for all integer $m < M$

$$\int_{-\infty}^{\infty} t^m \psi(t)\,dt = 0.$$

The mother wavelet is scaled (or dilated) by a factor of a and translated (or shifted) by a factor of b to give (under Morlet's original formulation):

$$\psi_{a,b}(t) = \frac{1}{\sqrt{a}}\psi\left(\frac{t-b}{a}\right).$$

For the continuous WT, the pair (a,b) varies over the full half-plane $R_+ \times R$; for the discrete WT this pair varies over a discrete subset of it, which is also called affine group.

These functions are often incorrectly referred to as the basis functions of the (continuous) transform. In fact, as in the continuous Fourier transform, there is no basis in the continuous wavelet transform. Time-frequency interpretation uses a subtly different formulation (after Delprat).

Restrictionÿ

(1) $\frac{1}{\sqrt{a}}\int_{-\infty}^{\infty}\varphi_{a1,b1}(t)\varphi\left(\frac{t-b}{a}\right)dt$ when $a1 = a$ and $b1 = b$,

(2) $\Psi(t)$ has a finite time interval

Comparisons with Fourier Transform (Continuous-Time)

The wavelet transform is often compared with the Fourier transform, in which signals are represented as a sum of sinusoids. In fact, the Fourier transform can be viewed as a special case of the continuous wavelet transform with the choice of the mother wavelet $\psi(t) = e^{-2\pi it}$.

The main difference in general is that wavelets are localized in both time and frequency whereas the standard Fourier transform is only localized in frequency. The Short-time Fourier transform (STFT) is similar to the wavelet transform, in that it is also time and frequency localized, but there are issues with the frequency/time resolution trade-off.

In particular, assuming a rectangular window region, one may think of the STFT as a transform with a slightly different kernel

$$\psi(t) = g(t-u)e^{-2\pi it}$$

where $g(t-u)$ can often be written as $\text{rect}\left(\frac{\text{t-u}}{\Delta_\text{t}}\right)$, where Δ_t and u respectively denote the length and temporal offset of the windowing function. Using Parseval's theorem, one may define the wavelet's energy as

$$E = \int_{-\infty}^{\infty}|\psi(t)|^2\,dt = \frac{1}{2\pi}\int_{-\infty}^{\infty}|\widehat{\psi(\omega)}|^2\,d\omega$$

From this, the square of the temporal support of the window offset by time u is given by

$$\sigma_t^2 = \frac{1}{E}\int|t-u|^2|\psi(t)|^2\,dt$$

and the square of the spectral support of the window acting on a frequency ξ

$$\sigma_\omega^2 = \frac{1}{2\pi E} \int |\omega - \xi|^2 |\hat{\psi}(\omega)|^2 d\omega$$

As stated by the Heisenberg uncertainty principle, the product of the temporal and spectral supports $\sigma_t^2 \sigma_\omega^2 \geq 1/4$ for any given time-frequency atom, or resolution cell. The STFT windows restrict the resolution cells to spectral and temporal supports determined by .

Multiplication with a rectangular window in the time domain corresponds to convolution with a $\text{sinc}(\Delta_t \omega)$ function in the frequency domain, resulting in spurious ringing artifacts for short/localized temporal windows. With the continuous-time Fourier Transform, $\Delta_t \to \infty$ and this convolution is with a delta function in Fourier space, resulting in the true Fourier transform of the signal $x(t)$. The window function may be some other apodizing filter, such as a Gaussian. The choice of windowing function will affect the approximation error relative to the true Fourier transform.

A given resolution cell's time-bandwidth product may not be exceeded with the STFT. All STFT basis elements maintain a uniform spectral and temporal support for all temporal shifts or offsets, thereby attaining an equal resolution in time for lower and higher frequencies. The resolution is purely determined by the sampling width.

In contrast, the wavelet transform's multiresolutional properties enables large temporal supports for lower frequencies while maintaining short temporal widths for higher frequencies by the scaling properties of the wavelet transform. This property extends conventional time-frequency analysis into time-scale analysis.

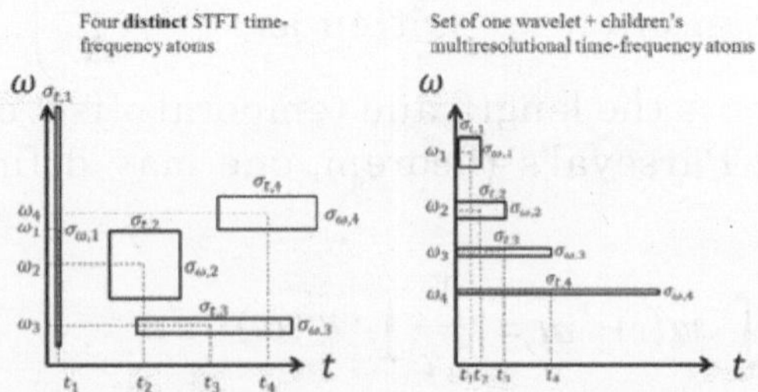

Figure: *STFT time-frequency atoms (left) and DWT time-scale atoms (right). The time-frequency atoms are four different basis functions used for the STFT (i.e. four separate Fourier transforms required). The time-scale atoms of the DWT achieve small temporal widths for high frequencies and good temporal widths for low frequencies with a single transform basis set.*

The discrete wavelet transform is less computationally complex, taking O(N) time as compared to O(N log N) for the fast Fourier transform. This computational advantage is not inherent to the transform, but reflects the choice of a logarithmic division of frequency, in contrast to the equally spaced frequency divisions of the FFT (Fast Fourier Transform) which uses the same basis functions as DFT (Discrete Fourier Transform).

It is also important to note that this complexity only applies when the filter size has no relation to the signal size. A wavelet without compact support such as the Shannon wavelet would require $O(N^2)$. (For instance, a logarithmic Fourier Transform also exists with O(N) complexity, but the original signal must be sampled logarithmically in time, which is only useful for certain types of signals.

Computational Methods

Computational statistics, or statistical computing, is the interface between statistics and computer science. It is the area of computational science (or scientific computing) specific to the mathematical science of statistics. This area is also developing rapidly, leading to calls that a broader concept of computing should be taught as part of general statistical education.

The terms 'computational statistics' and 'statistical computing' are often used interchangeably, although Carlo Lauro (a former president of the International Association for Statistical Computing) proposed making a distinction, defining 'statistical computing' as "the application of computer science to statistics", and 'computational statistics' as "aiming at the design of algorithm for implementing statistical methods on computers, including the ones unthinkable before the computer age (e.g. bootstrap, simulation), as well as to cope with analytically intractable problems"

The term 'Computational statistics' may also be used to refer to computationally intensive statistical methods including resampling methods, Markov chain Monte Carlo methods, local regression, kernel density estimation, artificial neural networks and generalized additive models.

Markov Chain Monte Carlo

In mathematics, more specifically in statistics, Markov chain Monte Carlo (MCMC) methods are a class of algorithms for sampling from a probability distribution based on constructing a Markov chain that has the desired distribution as its equilibrium distribution. The state of the chain after a number of steps is then used as a sample of the desired distribution. The quality of the sample improves as a function of the number of steps.

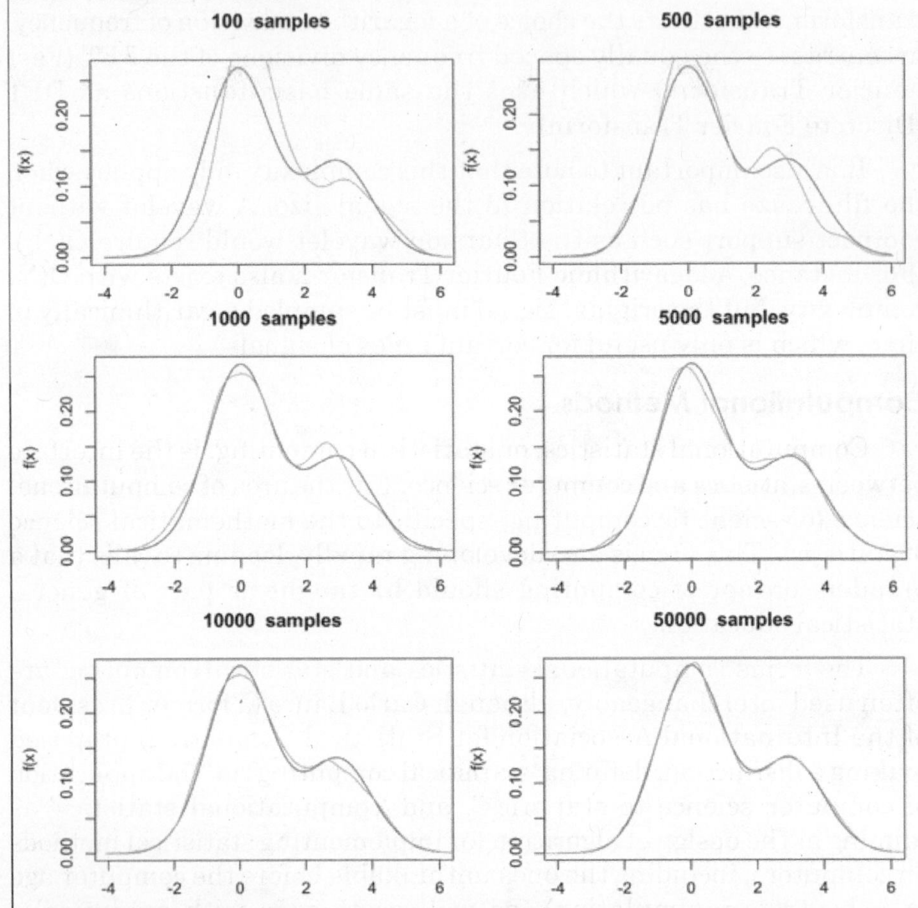

Figure: *Convergence of the Metropolis-Hastings algorithm. MCMC attempts to approximate the blue distribution with the orange distribution*

Random walk Monte Carlo methods make up a large subclass of MCMC methods.

Application Domains

- MCMC methods are primarily used for calculating numerical approximations of multi-dimensional integrals, for example in Bayesian statistics, computational physics, computational biology and computational linguistics.
- They are also used for generating samples that gradually populate the rare failure region in rare event sampling.

Classification

Multi-Dimensional Integrals

When an MCMC method is used for approximating a multi-dimensional integral, an ensemble of "walkers" move around randomly. At each point where a walker steps, the integrand value at that point is counted towards the integral. The walker then may make a number of tentative steps around the area, looking for a place with a reasonably high contribution to the integral to move into next.

Random walk Monte Carlo methods are a kind of random simulation or Monte Carlo method. However, whereas the random samples of the integrand used in a conventional Monte Carlo integration are statistically independent, those used in MCMC methods are correlated. A Markov chain is constructed in such a way as to have the integrand as its equilibrium distribution.

Examples

Examples of random walk Monte Carlo methods include the following:

- Metropolis–Hastings algorithm: This method generates a random walk using a proposal density and a method for rejecting some of the proposed moves.
- Gibbs sampling: This method requires all the conditional distributions of the target distribution to be sampled exactly. It is popular partly because it does not require any 'tuning'.
- Slice sampling: This method depends on the principle that one can sample from a distribution by sampling uniformly from the region under the plot of its density function. It alternates uniform sampling in the vertical direction with uniform sampling from the horizontal 'slice' defined by the current vertical position.
- Multiple-try Metropolis: This method is a variation of the Metropolis–Hastings algorithm that allows multiple trials at each point. By making it possible to take larger steps at each iteration, it helps address the curse of dimensionality.
- Reversible-jump: This method is a variant of the Metropolis–Hastings algorithm that allows proposals that change the dimensionality of the space. MCMC methods that change dimensionality have long been used in statistical physics applications, where for some problems a distribution that is a grand canonical ensemble is used (e.g., when the number of

molecules in a box is variable). But the reversible-jump variant is useful when doing MCMC or Gibbs sampling over nonparametric Bayesian models such as those involving the Dirichlet process or Chinese restaurant process, where the number of mixing components/clusters/etc. is automatically inferred from the data.

Other MCMC Methods

Reducing Correlation

More sophisticated methods use various ways of reducing the correlation between successive samples. These algorithms may be harder to implement, but they usually exhibit faster convergence (i.e. fewer steps for an accurate result).

Examples

Examples of non-random walk MCMC methods include the following:

- Hybrid Monte Carlo (HMC): Tries to avoid random walk behaviour by introducing an auxiliary momentum vector and implementing Hamiltonian dynamics, so the potential energy function is the target density. The momentum samples are discarded after sampling. The end result of Hybrid Monte Carlo is that proposals move across the sample space in larger steps; they are therefore less correlated and converge to the target distribution more rapidly.
- Some variations on slice sampling also avoid random walks.
- Langevin MCMC and other methods that rely on the gradient (and possibly second derivative) of the log posterior avoid random walks by making proposals that are more likely to be in the direction of higher probability density.

Convergence

Usually it is not hard to construct a Markov chain with the desired properties. The more difficult problem is to determine how many steps are needed to converge to the stationary distribution within an acceptable error. A good chain will have rapid mixing: the stationary distribution is reached quickly starting from an arbitrary position.

Typically, MCMC sampling can only approximate the target distribution, as there is always some residual effect of the starting position. More sophisticated MCMC-based algorithms such as coupling from the past can produce exact samples, at the cost of additional computation and an unbounded (though finite in expectation) running time.

Many random walk Monte Carlo methods move around the equilibrium distribution in relatively small steps, with no tendency for the steps to proceed in the same direction. These methods are easy to implement and analyze, but unfortunately it can take a long time for the walker to explore all of the space. The walker will often double back and cover ground already covered.

Chaos Theory

Chaos theory is a field of study in mathematics, with applications in several disciplines including meteorology, sociology, physics, engineering, economics, biology, and philosophy. Chaos theory studies the behaviour of dynamical systems that are highly sensitive to initial conditions—a response popularly referred to as the butterfly effect. Small differences in initial conditions (such as those due to rounding errors in numerical computation) yield widely diverging outcomes for such dynamical systems, rendering long-term prediction impossible in general. This happens even though these systems are deterministic, meaning that their future behaviour is fully determined by their initial conditions, with no random elements involved. In other words, the deterministic nature of these systems does not make them predictable. This behaviour is known as deterministic chaos, or simply chaos. The theory was summarized by Edward Lorenz as follows:

Chaos: When the present determines the future, but the approximate present does not approximately determine the future.

Chaotic behaviour can be observed in many natural systems, such as weather and climate. This behaviour can be studied through analysis of a chaotic mathematical model, or through analytical techniques such as recurrence plots and Poincaré maps.

Chaos theory concerns deterministic systems whose behaviour can in principle be predicted. Chaotic systems are predictable for a while and then appear to become random. The amount of time for which the behaviour of a chaotic system can be effectively predicted depends on three things: How much uncertainty we are willing to tolerate in the forecast; how accurately we are able to measure its current state; and a time scale depending on the dynamics of the system, called the Lyapunov time. Some examples of Lyapunov times are: chaotic electrical circuits, ~1 millisecond; weather systems, a couple of days (unproven); the solar system, 50 million years. In chaotic systems the uncertainty in a forecast increases exponentially with elapsed time. Hence doubling the forecast time squares the proportional uncertainty in the forecast. This means that in practice a meaningful prediction cannot be made over an interval of more than two or three times the Lyapunov time. When meaningful predictions cannot be made, the system appears to be random.

Chaotic Dynamics

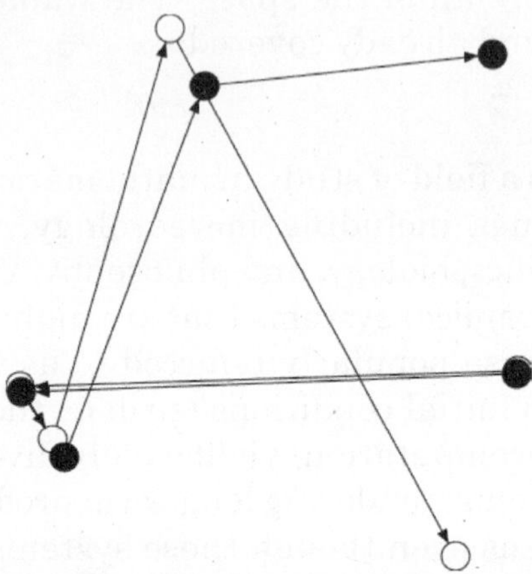

The map defined by $x \to 4x(1-x)$ and $y \to x + y \bmod 1$ displays sensitivity to initial conditions. Here two series of x and y values diverge markedly over time from a tiny initial difference.

In common usage, "chaos" means "a state of disorder". However, in chaos theory, the term is defined more precisely. Although there is no universally accepted mathematical definition of chaos, a commonly used definition says that, for a dynamical system to be classified as chaotic, it must have the following properties:

1. it must be sensitive to initial conditions;
2. it must be topologically mixing; and
3. it must have dense periodic orbits.

Sensitivity to Initial Conditions

Sensitivity to initial conditions means that each point in a chaotic system is arbitrarily closely approximated by other points with significantly different future paths, or trajectories. Thus, an arbitrarily small change, or perturbation, of the current trajectory may lead to significantly different future behaviour.

It has been shown that in some cases the last two properties in the above actually imply sensitivity to initial conditions, and if attention is restricted to intervals, the second property implies the other two (an alternative, and in general weaker, definition of chaos uses only the first two properties in the above list). It is interesting that the most practically significant property, that of sensitivity to initial conditions, is redundant in the definition, being implied by two (or for intervals, one) purely topological properties, which are therefore of greater interest to mathematicians.

Sensitivity to initial conditions is popularly known as the "butterfly effect", so called because of the title of a paper given by Edward Lorenz in 1972 to the American Association for the Advancement of Science in Washington, D.C., entitled Predictability: Does the Flap of a Butterfly's Wings in Brazil set off a Tornado in Texas?. The flapping wing represents a small change in the initial condition of the system, which causes a chain of events leading to large-scale phenomena. Had the butterfly not flapped its wings, the trajectory of the system might have been vastly different.

A consequence of sensitivity to initial conditions is that if we start with only a finite amount of information about the system (as is usually the case in practice), then beyond a certain time the system will no longer be predictable. This is most familiar in the case of weather, which is generally predictable only about a week ahead. Of course this does not mean that we cannot say anything about events far in the future; there are some restrictions on the system. With weather, we know that the temperature will never reach 100 degrees Celsius or fall to -130 degrees Celsius on earth, but we are not able to say exactly what day we will have the hottest temperature of the year.

In more mathematical terms, the Lyapunov exponent measures the sensitivity to initial conditions. Given two starting trajectories in the phase space that are infinitesimally close, with initial separation $\delta \mathbf{Z}_0$ end up diverging at a rate given by

$$|\delta \mathbf{Z}(t)| \approx e^{\lambda t} |\delta \mathbf{Z}_0|$$

where t is the time and λ is the Lyapunov exponent. The rate of separation depends on the orientation of the initial separation vector, so there is a whole spectrum of Lyapunov exponents. The number of Lyapunov exponents is equal to the number of dimensions of the phase space, though it is common to just refer to the largest one. For example, the maximal Lyapunov exponent (MLE) is most often used because it determines the overall predictability of the system. A positive MLE is usually taken as an indication that the system is chaotic.

There are also other properties that relate to sensitivity of initial conditions, such as measure-theoretical mixing (as discussed in ergodic theory) and properties of a K-system.

Topological Mixing

Topological mixing (or topological transitivity) means that the system will evolve over time so that any given region or open set of its phase space will eventually overlap with any other given region. This mathematical concept of "mixing" corresponds to the standard intuition, and the mixing of coloured dyes or fluids is an example of a chaotic system.

Topological mixing is often omitted from popular accounts of chaos, which equate chaos with only sensitivity to initial conditions. However, sensitive dependence on initial conditions alone does not give chaos. For example, consider the simple dynamical system produced by repeatedly doubling an initial value.

This system has sensitive dependence on initial conditions everywhere, since any pair of nearby points will eventually become widely separated. However, this example has no topological mixing, and therefore has no chaos. Indeed, it has extremely simple behaviour: all points except 0 will tend to positive or negative infinity.

Density of Periodic Orbits

For a chaotic system to have a dense periodic orbit means that every point in the space is approached arbitrarily closely by periodic orbits. The one-dimensional logistic map defined by $x \to 4x\,(1 - x)$ is one of the simplest systems with density of periodic orbits. For example, $\frac{5+\sqrt{5}}{8} \to \frac{5-\sqrt{5}}{8} \to$ (or approximately $0.3454915 \to 0.9045085 \to 0.3454915$) is an (unstable) orbit of period 2, and similar orbits exist for periods 4, 8, 16, etc. (indeed, for all the periods specified by Sharkovskii's theorem).

Sharkovskii's theorem is the basis of the Li and Yorke (1975) proof that any one-dimensional system that exhibits a regular cycle of period three will also display regular cycles of every other length as well as completely chaotic orbits.

Strange Attractors

Some dynamical systems, like the one-dimensional logistic map defined by $x \to 4x\,(1 - x)$, are chaotic everywhere, but in many cases chaotic behaviour is found only in a subset of phase space. The cases of most interest arise when the chaotic behaviour takes place on an attractor, since then a large set of initial conditions will lead to orbits that converge to this chaotic region.

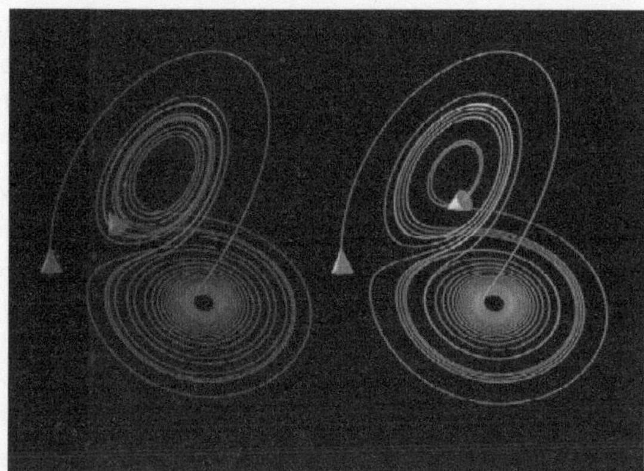

Figure: *The Lorenz attractor displays chaotic behaviour. These two plots demonstrate sensitive dependence on initial conditions within the region of phase space occupied by the attractor.*

An easy way to visualize a chaotic attractor is to start with a point in the basin of attraction of the attractor, and then simply plot its subsequent orbit. Because of the topological transitivity condition, this is likely to produce a picture of the entire final attractor, and indeed both orbits shown in the figure on the right give a picture of the general shape of the Lorenz attractor.

This attractor results from a simple three-dimensional model of the Lorenz weather system. The Lorenz attractor is perhaps one of the best-known chaotic system diagrams, probably because it was not only one of the first, but it is also one of the most complex and as such gives rise to a very interesting pattern, that with a little imagination, looks like the wings of a butterfly.

Unlike fixed-point attractors and limit cycles, the attractors that arise from chaotic systems, known as strange attractors, have great detail and complexity. Strange attractors occur in both continuous dynamical systems (such as the Lorenz system) and in some discrete systems (such as the Hénon map). Other discrete dynamical systems have a repelling structure called a Julia set which forms at the boundary between basins of attraction of fixed points – Julia sets can be thought of as strange repellers. Both strange attractors and Julia sets typically have a fractal structure, and the fractal dimension can be calculated for them.

Minimum Complexity of a Chaotic System

Bifurcation diagram of the logistic map $x \to r\,x\,(1-x)$. Each vertical slice shows the attractor for a specific value of r. The diagram displays period-doubling as r increases, eventually producing chaos.

Discrete chaotic systems, such as the logistic map, can exhibit strange attractors whatever their dimensionality. In contrast, for continuous dynamical systems, the Poincaré–Bendixson theorem shows that a strange attractor can only arise in three or more dimensions. Finite-dimensional linear systems are never chaotic; for a dynamical system to display chaotic behaviour, it has to be either nonlinear or infinite-dimensional.

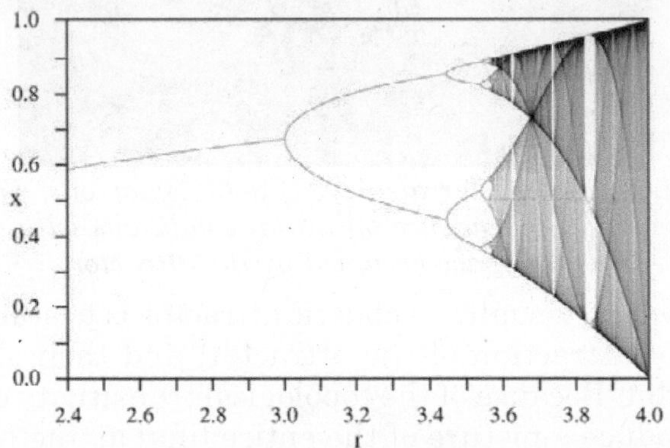

The Poincaré–Bendixson theorem states that a two-dimensional differential equation has very regular behaviour. The Lorenz attractor discussed above is generated by a system of three differential equations such as:

$$\frac{dx}{dt} = \sigma y - \sigma x,$$

$$\frac{dy}{dt} = \rho x - xz - y,$$

$$\frac{dz}{dt} = xy - \beta z.$$

where x, y, and z make up the system state, t is time, and σ, ρ, $\beta$are the system parameters. Five of the terms on the right hand side are linear, while two are quadratic; a total of seven terms.

Another well-known chaotic attractor is generated by the Rossler equations which have only one nonlinear term out of seven. Sprott found a three-dimensional system with just five terms, that had only one nonlinear term, which exhibits chaos for certain parameter values. Zhang and Heidel showed that, at least for dissipative and conservative quadratic systems, three-dimensional quadratic systems with only three or four terms on the right-hand side cannot exhibit chaotic behaviour. The reason is, simply put, that solutions to such systems are asymptotic to a two-dimensional surface and therefore solutions are well behaved.

While the Poincaré–Bendixson theorem shows that a continuous dynamical system on the Euclidean plane cannot be chaotic, two-dimensional continuous systems with non-Euclidean geometry can exhibit chaotic behaviour. Perhaps surprisingly, chaos may occur also in linear systems, provided they are infinite dimensional. A theory of linear chaos is being developed in a branch of mathematical analysis known as functional analysis.

Jerk Systems

In physics, jerk is the third derivative of position, and such, in mathematics differential equations of the form

$$J\left(\dddot{x}, \ddot{x}, \dot{x}, x\right) = 0$$

are sometimes called Jerk equations. It has been shown, that a jerk equation, which is equivalent to a system of three first order, ordinary, non-linear differential equation is in a certain sense the minimal setting for solutions showing chaotic behaviour. This motivates mathematical interest in jerk systems. Systems involving a fourth or higher derivative are called accordingly hyperjerk systems.

A jerk system is a system whose behaviour is described by a jerk equation, and for certain jerk equations simple electronic circuits may be designed which model the solutions to this equation. These circuits are known as jerk circuits.

One of the most interesting properties of jerk circuits is the possibility of chaotic behaviour. In fact, certain well-known chaotic systems, such as the Lorenz attractor and the Rφssler map, are conventionally described as a system of three first-order differential equations, but which may be combined into a single (although rather complicated) jerk equation. It has been shown, that non-linear jerk systems are in a sense minimally complex systems to show chaotic behaviour, there is no chaotic system involving only two first order, ordinary differential equations (the system resulting in an equation of second order only).

An example of a jerk equation with non-linearity in the magnitude of x , is:

$$\frac{d^3x}{dt^3} + A\frac{d^2x}{dt^2} + \frac{dx}{dt} - |x| + 1 = 0.$$

Here A is an adjustable parameter. This equation has a chaotic solution for A=3/5 and can be implemented with the following jerk circuit; the required non-linearity is brought about by the two diodes:

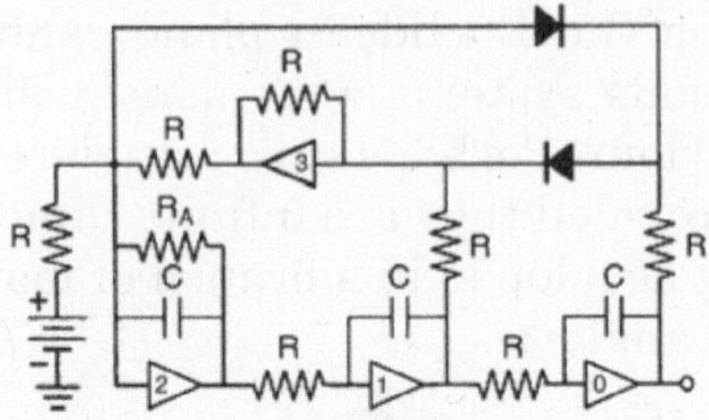

In the above circuit, all resistors are of equal value, except $R_A = R / A = 5R / 3$, and all capacitors are of equal size. The dominant frequency will be $1/2\pi RC$. The output of op amp 0 will correspond to the x variable, the output of 1 will correspond to the first derivative of x and the output of 2 will correspond to the second derivative.

Spontaneous Order

Under the right conditions chaos will spontaneously evolve into a lockstep pattern. In the Kuramoto model, four conditions suffice to produce synchronization in a chaotic system. Examples include the coupled oscillation of Christiaan Huygens' pendulums, fireflies, neurons, the London Millenium Bridge resonance, and large arrays of Josephson junctions.

Chapter 7

Principal Component Analysis

Principal component analysis (PCA) is a statistical procedure that uses an orthogonal transformation to convert a set of observations of possibly correlated variables into a set of values of linearly uncorrelated variables called principal components.

The number of principal components is less than or equal to the number of original variables. This transformation is defined in such a way that the first principal component has the largest possible variance (that is, accounts for as much of the variability in the data as possible), and each succeeding component in turn has the highest variance possible under the constraint that it is orthogonal to (i.e., uncorrelated with) the preceding components. The principal components are orthogonal because they are the eigenvectors of the covariance matrix, which is symmetric. PCA is sensitive to the relative scaling of the original variables.

Depending on the field of application, it is also named the discrete Karhunen–Loθve transform (KLT) in signal processing, the Hotelling transform in multivariate quality control, proper orthogonal decomposition (POD) in mechanical engineering, singular value decomposition (SVD) of X (Golub and Van Loan, 1983), eigenvalue decomposition (EVD) of X^TX in linear algebra, factor analysis (for a discussion of the differences between PCA and factor analysis), Eckart–Young theorem (Harman, 1960), or Schmidt–Mirsky theorem in psychometrics, empirical orthogonal functions (EOF) in meteorological science, empirical eigenfunction decomposition (Sirovich, 1987), empirical component analysis (Lorenz, 1956), quasiharmonic modes (Brooks et al., 1988), spectral decomposition in noise and vibration, and empirical modal analysis in structural dynamics.

PCA was invented in 1901 by Karl Pearson, as an analogue of the principal axes theorem in mechanics; it was later independently developed (and named) by Harold Hotelling in the 1930s. The method is mostly used as a tool in exploratory data analysis and for making predictive models. PCA can be done by eigenvalue decomposition of a data covariance (or correlation) matrix or singular value decomposition of a data matrix, usually after mean centring (and normalizing or using Z-scores) the data matrix for each attribute. The results of a PCA are usually discussed in terms of component scores, sometimes called factor scores (the transformed variable values corresponding to a particular data point), and loadings (the weight by which each standardized original variable should be multiplied to get the component score).

PCA is the simplest of the true eigenvector-based multivariate analyses. Often, its operation can be thought of as revealing the internal structure of the data in a way that best explains the variance in the data. If a multivariate dataset is visualised as a set of coordinates in a high-dimensional data space (1 axis per variable), PCA can supply the user with a lower-dimensional picture, a projection or "shadow" of this object when viewed from its (in some sense) most informative viewpoint. This is done by using only the first few principal components so that the dimensionality of the transformed data is reduced.

PCA is closely related to factor analysis. Factor analysis typically incorporates more domain specific assumptions about the underlying structure and solves eigenvectors of a slightly different matrix.

PCA is also related to canonical correlation analysis (CCA). CCA defines coordinate systems that optimally describe the cross-covariance between two datasets while PCA defines a new orthogonal coordinate system that optimally describes variance in a single dataset.

Intuition

PCA can be thought of as fitting an n-dimensional ellipsoid to the data, where each axis of the ellipsoid represents a principal component. If some axis of the ellipse is small, then the variance along that axis is also small, and by omitting that axis and its corresponding principal component from our representation of the dataset, we lose only a commensurately small amount of information.

To find the axes of the ellipse, we must first subtract the mean of each variable from the dataset to centre the data around the origin. Then, we compute the covariance matrix of the data, and calculate the eigenvalues and their corresponding eigenvectors of this covariance matrix. Then, we must orthogonalize the set of eigenvectors, and normalize each to become unit vectors. Once this is done, each of the mutually orthogonal, unit eigenvectors can be interpreted as an axis of the ellipsoid fitted to the data. The proportion of the variance that each eigenvector represents can be calculated by dividing the eigenvalue corresponding to that eigenvector by the sum of all eigenvalues.

It is important to note that this procedure is sensitive to the scaling of the data, and that there is no consensus as to how to best scale the data to obtain optimal results.

Details

PCA is mathematically defined as an orthogonal linear transformation that transforms the data to a new coordinate system such that the greatest variance by some projection of the data comes to lie on the first coordinate (called the first principal component), the second greatest variance on the second coordinate, and so on.

Consider a data matrix, X, with column-wise zero empirical mean (the sample mean of each column has been shifted to zero), where each of the n rows represents a different repetition of the experiment, and each of the p columns gives a particular kind of datum (say, the results from a particular sensor).

Mathematically, the transformation is defined by a set of p-dimensional vectors of weights or loadings $\mathbf{w}_{(k)} = (w_1, \ldots, w_p)_{(k)}$ that map each row vector $\mathbf{x}_{(i)}$ of X to a new vector of principal component scores $\mathbf{t}_{(i)} = (t_1, \ldots, t_p)_{(i)}$, given by

$$t_{k(i)} = \mathbf{x}_{(i)} \cdot \mathbf{w}_{(k)}$$

in such a way that the individual variables of t considered over the data set successively inherit the maximum possible variance from x, with each loading vector w constrained to be a unit vector.

First Component

The first loading vector $\mathrm{w}_{(1)}$ thus has to satisfy

$$\mathbf{w}_{(1)} = \underset{\|\mathbf{w}\|=1}{\operatorname{argmax}}\{\sum_i (t_1)_{(i)}^2\} = \underset{\|\mathbf{w}\|=1}{\operatorname{argmax}} \sum_i \left(\mathbf{x}_{(i)} \cdot \mathbf{w}\right)^2$$

Equivalently, writing this in matrix form gives

$$\mathbf{w}_{(1)} = \underset{\|\mathbf{w}\|=1}{\operatorname{argmax}}\{\| \mathbf{X}\mathbf{w} \|^2\} = \underset{\|\mathbf{w}\|=1}{\operatorname{argmax}}\{\mathbf{w}^T \mathbf{X}^T \mathbf{X}\mathbf{w}\}$$

Since $w_{(1)}$ has been defined to be a unit vector, it equivalently also satisfies

$$\mathbf{w}_{(1)} = \arg\max\left\{\frac{\mathbf{w}^T\mathbf{X}^T\mathbf{X}\mathbf{w}}{\mathbf{w}^T\mathbf{w}}\right\}$$

The quantity to be maximised can be recognised as a Rayleigh quotient. A standard result for a symmetric matrix such as X^TX is that the quotient's maximum possible value is the largest eigenvalue of the matrix, which occurs when w is the corresponding eigenvector.

With $w_{(1)}$ found, the first component of a data vector $x_{(i)}$ can then be given as a score $t_{1(i)} = x_{(i)} \cdot w_{(1)}$ in the transformed co-ordinates, or as the corresponding vector in the original variables, $\{x_{(i)} \cdot w_{(1)}\}\, w_{(1)}$.

Further Components

The kth component can be found by subtracting the first $k-1$ principal components from X:

$$\hat{\mathbf{X}}_k = \mathbf{X} - \sum_{s=1}^{k-1} \mathbf{X}\mathbf{w}_{(s)}\mathbf{w}_{(s)}^{\mathrm{T}}$$

and then finding the loading vector which extracts the maximum variance from this new data matrix

$$\mathbf{w}_{(k)} = \underset{\|\mathbf{w}\|=1}{\arg\max}\left\{\|\hat{\mathbf{X}}_{k-1}\mathbf{w}\|^2\right\} = \arg\max\left\{\frac{\mathbf{w}^T\hat{\mathbf{X}}_{k-1}^T\hat{\mathbf{X}}_{k-1}\mathbf{w}}{\mathbf{w}^T\mathbf{w}}\right\}$$

It turns out that this gives the remaining eigenvectors of X^TX, with the maximum values for the quantity in brackets given by their corresponding eigenvalues.

The kth principal component of a data vector $\mathrm{x}_{(\mathrm{i})}$ can therefore be given as a score $t_{k(i)} = x_{(i)} \cdot w_{(k)}$ in the transformed co-ordinates, or as the corresponding vector in the space of the original variables, $\{x_{(i)} \cdot w_{(k)}\}$ $w_{(k)}$, where $w_{(k)}$ is the kth eigenvector of X^TX.

The full principal components decomposition of X can therefore be given as

$$\mathbf{T} = \mathbf{X}\mathbf{W}$$

where W is a p-by-p matrix whose columns are the eigenvectors of X^TX

Covariances

$\mathrm{X}^\mathrm{T}\mathrm{X}$ itself can be recognised as proportional to the empirical sample covariance matrix of the dataset X.

The sample covariance Q between two of the different principal components over the dataset is given by:

$$
\begin{aligned}
Q(\mathrm{PC}_{(j)},\mathrm{PC}_{(k)}) &\propto (\mathbf{Xw}_{(j)})\cdot(\mathbf{Xw}_{(k)}) \\
&= \mathbf{w}_{(j)}^T\mathbf{X}^T\mathbf{X}\mathbf{w}_{(k)} \\
&= \mathbf{w}_{(j)}^T\lambda_{(k)}\mathbf{w}_{(k)} \\
&= \lambda_{(k)}\mathbf{w}_{(j)}^T\mathbf{w}_{(k)}
\end{aligned}
$$

where the eigenvalue property of $w_{(k)}$ has been used to move from line 2 to line 3. However eigenvectors $w_{(j)}$ and $w_{(k)}$ corresponding to eigenvalues of a symmetric matrix are orthogonal (if the eigenvalues are different), or can be orthogonalised (if the vectors happen to share an equal repeated value). The product in the final line is therefore zero; there is no sample covariance between different principal components over the dataset.

Another way to characterise the principal components transformation is therefore as the transformation to coordinates which diagonalise the empirical sample covariance matrix.

In matrix form, the empirical covariance matrix for the original variables can be written

$$\mathbf{Q} \propto \mathbf{X}^T\mathbf{X} = \mathbf{W}\mathbf{\Lambda}\mathbf{W}^T$$

The empirical covariance matrix between the principal components becomes

$$\mathbf{W}^T\mathbf{Q}\mathbf{W} \propto \mathbf{W}^T\mathbf{W}\mathbf{\Lambda}\mathbf{W}^T\mathbf{W} = \mathbf{\Lambda}$$

where Λ is the diagonal matrix of eigenvalues $\lambda_{(k)}$ of X^TX ($\lambda_{(k)}$ being equal to the sum of the squares over the dataset associated with each component k: $\lambda_{(k)} = \Sigma_i t_{k\,(i)}^2 = \Sigma_i (x_{(i)} E\!- w_{(k)})^2$)

Dimensionality Reduction

The faithful transformation $T = X\,W$ maps a data vector $x_{(i)}$ from an original space of p variables to a new space of p variables which are uncorrelated over the dataset. However, not all the principal components need to be kept. Keeping only the first L principal components, produced by using only the first L loading vectors, gives the truncated transformation

$$\mathbf{T}_L = \mathbf{X}\mathbf{W}_L$$

where the matrix T_L now has n rows but only L columns. In other words, PCA learns a linear transformation $t = W^T x, x \in R^p, t \in R^L$, where the columns of $p \times L$ matrix W form an orthogonal basis for the L features (the components of representation t) that are decorrelated. By construction, of all the transformed data matrices with only L

columns, this score matrix maximises the variance in the original data that has been preserved, while minimising the total squared reconstruction error $\| \mathbf{TW}^T - \mathbf{T}_L \mathbf{W}_L^T \|_2^2$ or $\| \mathbf{X} - \mathbf{X}_L \|_2^2$.

Such dimensionality reduction can be a very useful step for visualising and processing high-dimensional datasets, while still retaining as much of the variance in the dataset as possible. For example, selecting $L = 2$ and keeping only the first two principal components finds the two-dimensional plane through the high-dimensional dataset in which the data is most spread out, so if the data contains clusters these too may be most spread out, and therefore most visible to be plotted out in a two-dimensional diagram; whereas if two directions through the data (or two of the original variables) are chosen at random, the clusters may be much less spread apart from each other, and may in fact be much more likely to substantially overlay each other, making them indistinguishable.

Similarly, in regression analysis, the larger the number of explanatory variables allowed, the greater is the chance of overfitting the model, producing conclusions that fail to generalise to other datasets. One approach, especially when there are strong correlations between different possible explanatory variables, is to reduce them to a few principal components and then run the regression against them, a method called principal component regression.

Dimensionality reduction may also be appropriate when the variables in a dataset are noisy. If each column of the dataset contains independent identically distributed Gaussian noise, then the columns of T will also contain similarly identically distributed Gaussian noise (such a distribution is invariant under the effects of the matrix W, which can be thought of as a high-dimensional rotation of the co-ordinate axes). However, with more of the total variance concentrated in the first few principal components compared to the same noise variance, the proportionate effect of the noise is less—the first few components achieve a higher signal-to-noise ratio. PCA thus can have the effect of concentrating much of the signal into the first few principal components, which can usefully be captured by dimensionality reduction; while the later principal components may be dominated by noise, and so disposed of without great loss.

Singular Value Decomposition

The principal components transformation can also be associated with another matrix factorisation, the singular value decomposition (SVD) of X,

$$\mathbf{X} = \mathbf{U}\boldsymbol{\Sigma}\mathbf{W}^T$$

Here Σ is a n-by-p rectangular diagonal matrix of positive numbers $\sigma_{(k)}$, called the singular values of *X*; *U* is an n-by-n matrix, the columns of which are orthogonal unit vectors of length n called the left singular vectors of *X*; and W is a *p*-by-*p* whose columns are orthogonal unit vectors of length p and called the right singular vectors of *X*.

In terms of this factorisation, the matrix X^TX can be written

$$\begin{aligned}\mathbf{X}^T\mathbf{X} &= \mathbf{W}\boldsymbol{\Sigma}\mathbf{U}^T\mathbf{U}\boldsymbol{\Sigma}\mathbf{W}^T \\ &= \mathbf{W}\boldsymbol{\Sigma}^2\mathbf{W}^T\end{aligned}$$

Comparison with the eigenvector factorisation of X^TX establishes that the right singular vectors W of X are equivalent to the eigenvectors of X^TX, while the singular values $\sigma_{(k)}$ of X are equal to the square roots of the eigenvalues $\lambda_{(k)}$ of X^TX.

Using the singular value decomposition the score matrix *T* can be written

$$\begin{aligned}\mathbf{T} &= \mathbf{X}\mathbf{W} \\ &= \mathbf{U}\boldsymbol{\Sigma}\mathbf{W}^T\mathbf{W} \\ &= \mathbf{U}\boldsymbol{\Sigma}\end{aligned}$$

so each column of *T* is given by one of the left singular vectors of *X* multiplied by the corresponding singular value.

Efficient algorithms exist to calculate the SVD of X without having to form the matrix X^TX, so computing the SVD is now the standard way to calculate a principal components analysis from a data matrix, unless only a handful of components are required.

As with the eigen-decomposition, a truncated *n*-by-*L* score matrix T_L can be obtained by considering only the first *L* largest singular values and their singular vectors:

$$\mathbf{T}_L = \mathbf{U}_L\boldsymbol{\Sigma}_L = \mathbf{X}\mathbf{W}_L$$

The truncation of a matrix *M* or *T* using a truncated singular value decomposition in this way produces a truncated matrix that is the nearest possible matrix of rank *L* to the original matrix, in the sense of the difference between the two having the smallest possible Frobenius norm, a result known as the Eckart–Young theorem [1936].

Further Considerations

Given a set of points in Euclidean space, the first principal component corresponds to a line that passes through the multidimensional mean and minimizes the sum of squares of the distances of the points from the line. The second principal component corresponds to the same concept after all correlation with the first principal component has been subtracted from the points. The singular values (in Σ) are the square roots of the eigenvalues of the matrix X^TX. Each eigenvalue is proportional to the portion of the "variance" (more correctly of the sum of the squared distances of the points from their multidimensional mean) that is correlated with each eigenvector.

The sum of all the eigenvalues is equal to the sum of the squared distances of the points from their multidimensional mean. PCA essentially rotates the set of points around their mean in order to align with the principal components. This moves as much of the variance as possible (using an orthogonal transformation) into the first few dimensions. The values in the remaining dimensions, therefore, tend to be small and may be dropped with minimal loss of information. PCA is often used in this manner for dimensionality reduction. PCA has the distinction of being the optimal orthogonal transformation for keeping the subspace that has largest "variance" (as defined above). This advantage, however, comes at the price of greater computational requirements if compared, for example and when applicable, to the discrete cosine transform, and in particular to the DCT-II which is simply known as the "DCT". Nonlinear dimensionality reduction techniques tend to be more computationally demanding than PCA.

PCA is sensitive to the scaling of the variables. If we have just two variables and they have the same sample variance and are positively correlated, then the PCA will entail a rotation by 45° and the "loadings" for the two variables with respect to the principal component will be equal. But if we multiply all values of the first variable by 100, then the first principal component will be almost the same as that variable, with a small contribution from the other variable, whereas the second component will be almost aligned with the second original variable. This means that whenever the different variables have different units (like temperature and mass), PCA is a somewhat arbitrary method of analysis. (Different results would be obtained if one used Fahrenheit rather than Celsius for example.) Note that Pearson's original paper was entitled "On Lines and Planes of Closest Fit to Systems of Points in Space" – "in space" implies physical Euclidean space where such concerns do not arise.

One way of making the PCA less arbitrary is to use variables scaled so as to have unit variance, by standardizing the data and hence use the autocorrelation matrix instead of the autocovariance matrix as a basis for PCA. However, this compresses the fluctuations in all dimensions of the signal space to unit variance.

Mean subtraction (a.k.a. "mean centring") is necessary for performing PCA to ensure that the first principal component describes the direction of maximum variance. If mean subtraction is not performed, the first principal component might instead correspond more or less to the mean of the data. A mean of zero is needed for finding a basis that minimizes the mean square error of the approximation of the data.

PCA is equivalent to empirical orthogonal functions (EOF), a name which is used in meteorology.

An autoencoder neural network with a linear hidden layer is similar to PCA. Upon convergence, the weight vectors of the K neurons in the hidden layer will form a basis for the space spanned by the first K principal components. Unlike PCA, this technique will not necessarily produce orthogonal vectors.

PCA is a popular primary technique in pattern recognition. It is not, however, optimized for class separability. An alternative is the linear discriminant analysis, which does take this into account.

Another application of PCA is reducing the number of parameters in the process of generating computational models of oil reservoirs.

Limitations

As noted above, the results of PCA depend on the scaling of the variables. A scale-invariant form of PCA has been developed.

The applicability of PCA is limited by certain assumptions made in its derivation.

PCA and Information Theory

The claim that the PCA used for dimensionality reduction preserves most of the information of the data is misleading. Indeed, without any assumption on the signal model, PCA cannot help to reduce the amount of information lost during dimensionality reduction, where information was measured using Shannon entropy.

Under the assumption that

$$\mathbf{x} = \mathbf{s} + \mathbf{n}$$

i.e., that the data vector x is the sum of the desired information-bearing signal s and a noise signal n one can show that PCA can be optimal for dimensionality reduction also from an information-theoretic point-of-view.

In particular, Linsker showed that if is Gaussian and is Gaussian noise with a covariance matrix proportional to the identity matrix, the PCA maximizes the mutual information between the desired information and the dimensionality-reduced output .

If the noise is still Gaussian and has a covariance matrix proportional to the identity matrix (i.e., the components of the vector are iid), but the information-bearing signal is non-Gaussian (which is a common scenario), PCA at least minimizes an upper bound on the information loss, which is defined as

$$\mathbf{y} = \mathbf{W}_L^T \mathbf{x}$$

The optimality of PCA is also preserved if the noise is iid and at least more Gaussian (in terms of the Kullback–Leibler divergence) than the information-bearing signal .

In general, even if the above signal model holds, PCA loses its information-theoretic optimality as soon as the noise becomes dependent.

Computing PCA Using the Covariance Method

The following is a detailed description of PCA using the covariance method. But note that it is better to use the singular value decomposition (using standard software).

The goal is to transform a given data set X of dimension p to an alternative data set Y of smaller dimension L. Equivalently, we are seeking to find the matrix Y, where Y is the Karhunen–Loève transform (KLT) of matrix X:

$$\mathbf{Y} = \mathbb{KLT}\{\mathbf{X}\}$$

Organise the Data Set

Suppose you have data comprising a set of observations of p variables, and you want to reduce the data so that each observation can be described with only L variables, $L < p$. Suppose further, that the data are arranged as a set of n data vectors with each representing a single grouped observation of the p variables.

- Write as row vectors, each of which has p columns.
- Place the row vectors into a single matrix X of dimensions $n \times p$.

Derivation of PCA Using the Covariance Method

Let X be a d-dimensional random vector expressed as column vector. Without loss of generality, assume X has zero mean.

We want to find (*) a $d \times d$ orthonormal transformation matrix P so that PX has a diagonal covariant matrix (i.e. PX is a random vector with all its distinct components pairwise uncorrelated).

A quick computation assuming P were unitary yields:

$$t]rcl\,\text{var}(PX) = \mathbb{E}[PX\ (PX)^{\dagger}]$$
$$= \mathbb{E}[PX\ X^{\dagger}P^{\dagger}]$$
$$= P\,\mathbb{E}[XX^{\dagger}]P^{\dagger}$$
$$= P\ \text{var}(X)P^{-1}$$

Hence (*) holds if and only if $\text{var}(X)$ were diagonalisable by P.

This is very constructive, as var(X) is guaranteed to be a non-negative definite matrix and thus is guaranteed to be diagonalisable by some unitary matrix.

Iterative Computation

In practical implementations especially with high dimensional data (large p), the covariance method is rarely used because it is not efficient. One way to compute the first principal component efficiently is shown in the following pseudo-code, for a data matrix X with zero mean, without ever computing its covariance matrix.

$\mathbf{r} =$ a random vector of length p

do c times:

$\mathbf{s} = 0$ (a vector of length p)

for each row $\mathbf{x} \in \mathbf{X}$

$\mathbf{s} = \mathbf{s} + (\mathbf{x}\cdot\mathbf{r})\mathbf{x}$

$$\mathbf{r} = \frac{\mathbf{s}}{|\mathbf{s}|}$$

return r

This algorithm is simply an efficient way of calculating $X^T X\ r$, normalizing, and placing the result back in r (power iteration). It avoids the np^2 operations of calculating the covariance matrix. r will typically get close to the first principal component of X within a small number of iterations, c. (The magnitude of s will be larger after each iteration. Convergence can be detected when it increases by an amount too small for the precision of the machine.)

Subsequent principal components can be computed by subtracting component r from X and then repeating this algorithm to find the next principal component. However this simple approach is not numerically stable if more than a small number of principal components are required, because imprecisions in the calculations will additively affect the estimates of subsequent principal components. More advanced methods build on this basic idea, as with the closely related Lanczos algorithm.

One way to compute the eigenvalue that corresponds with each principal component is to measure the difference in mean-squared-distance between the rows and the centroid, before and after subtracting out the principal component. The eigenvalue that corresponds with the component that was removed is equal to this difference.

The NIPALS Method

For very-high-dimensional datasets, such as those generated in the *omics sciences (e.g., genomics, metabolomics) it is usually only necessary to compute the first few PCs. The non-linear iterative partial least squares (NIPALS) algorithm calculates t_1 and w_1^T from X. The outer product, $t_1 w_1^T$ can then be subtracted from X leaving the residual matrix E_1. This can be then used to calculate subsequent PCs. This results in a dramatic reduction in computational time since calculation of the covariance matrix is avoided.

However, for large data matrices, or matrices that have a high degree of column collinearity, NIPALS suffers from loss of orthogonality due to machine precision limitations accumulated in each iteration step. A Gram–Schmidt (GS) re-orthogonalization algorithm is applied to both the scores and the loadings at each iteration step to eliminate this loss of orthogonality.

Online/Sequential Estimation

In an "online" or "streaming" situation with data arriving piece by piece rather than being stored in a single batch, it is useful to make an estimate of the PCA projection that can be updated sequentially. This can be done efficiently, but requires different algorithms.

PCA and Qualitative Variables

In PCA, it is common that we want to introduce qualitative variables as supplementary elements. For example, many quantitative variables have been measured on plants. For these plants, some qualitative variables are available as, for example, the species to which the plant belongs. These data were subjected to PCA for quantitative variables. When analyzing the results, it is natural to connect the principal components to the qualitative variable species. For this, the following results are produced.

- Identification, on the factorial planes, of the different species e.g. using different colours.
- Representation, on the factorial planes, of the centres of gravity of plants belonging to the same species.
- For each centre of gravity and each axis, p-value to judge the significance of the difference between the centre of gravity and origin.

These results are what is called introducing a qualitative variable as supplementary element. This procedure is detailed in and Husson, Lκ & Pagès 2009 and Pagès 2013. Few software offer this option in an "automatic" way. This is the case of SPAD that historically, following the work of Ludovic Lebart, was the first to propose this option, and the R package FactoMineR.

Applications

Neuroscience

A variant of principal components analysis is used in neuroscience to identify the specific properties of a stimulus that increase a neuron's probability of generating an action potential. This technique is known as spike-triggered covariance analysis. In a typical application an experimenter presents a white noisc process as a stimulus (usually either as a sensory input to a test subject, or as a current injected directly into the neuron) and records a train of action potentials, or spikes, produced by the neuron as a result. Presumably, certain features of the stimulus make the neuron more likely to spike. In order to extract these features, the experimenter calculates the covariance matrix of the spike-triggered ensemble, the set of all stimuli (defined and discretized over a finite time window, typically on the order of 100 ms) that immediately preceded a spike.

The eigenvectors of the difference between the spike-triggered covariance matrix and the covariance matrix of the prior stimulus ensemble (the set of all stimuli, defined over the same length time window) then indicate the directions in the space of stimuli along which the variance of the spike-triggered ensemble differed the most from that of the prior stimulus ensemble. Specifically, the eigenvectors with the largest positive eigenvalues correspond to the directions along which the variance of the spike-triggered ensemble showed the largest positive change compared to the variance of the prior. Since these were the directions in which varying the stimulus led to a spike, they are often good approximations of the sought after relevant stimulus features.

In neuroscience, PCA is also used to discern the identity of a neuron from the shape of its action potential. Spike sorting is an important procedure because extracellular recording techniques often pick up signals from more than one neuron. In spike sorting, one first uses PCA to reduce the dimensionality of the space of action potential waveforms, and then performs clustering analysis to associate specific action potentials with individual neurons.

Relation between PCA and K-Means Clustering

It has been shown recently (2001,2004) that the relaxed solution of K-means clustering, specified by the cluster indicators, is given by the PCA principal components, and the PCA subspace spanned by the principal directions is identical to the cluster centroid subspace specified by the between-class scatter matrix. Thus PCA automatically projects to the subspace where the global solution of K-means clustering lies, and thus facilitates K-means clustering to find near-optimal solutions.

Relation between PCA and Factor Analysis

Principal components creates variables that are linear combinations of the original variables. The new variables have the property that the variables are all orthogonal. The principal components can be used to find clusters in a set of data. PCA is a variance-focused approach seeking to reproduce the total variable variance, in which components reflect both common and unique variance of the variable. PCA is generally preferred for purposes of data reduction (i.e., translating variable space into optimal factor space) but not when detect the latent construct or factors.

Factor analysis is similar to principal component analysis, in that factor analysis also involves linear combinations of variables. Different from PCA, factor analysis is a correlation-focused approach seeking to reproduce the inter-correlations among variables, in which the factors "represent the common variance of variables, excluding unique variance" . Factor analysis is generally used when the research purpose is detecting data structure (i.e., latent constructs or factors) or causal modelling.

Correspondence Analysis

Correspondence analysis (CA) was developed by Jean-Paul Benzécri and is conceptually similar to PCA, but scales the data (which should be non-negative) so that rows and columns are treated equivalently. It is traditionally applied to contingency tables. CA decomposes the chi-squared statistic associated to this table into orthogonal factors. Because CA is a descriptive technique, it can be applied to tables for which the chi-squared statistic is appropriate or not. Several variants of CA are available including detrended correspondence analysis and canonical correspondence analysis. One special extension is multiple correspondence analysis, which may be seen as the counterpart of principal component analysis for categorical data.

Generalizations

Nonlinear Generalizations

Most of the modern methods for nonlinear dimensionality reduction find their theoretical and algorithmic roots in PCA or K-means. Pearson's original idea was to take a straight line (or plane) which will be "the best fit" to a set of data points. Principal curves and manifolds give the natural geometric framework for PCA generalization and extend the geometric interpretation of PCA by explicitly constructing an embedded manifold for data approximation, and by encoding using standard geometric projection onto the manifold, as it is illustrated by Fig. Another popular generalization is kernel PCA, which corresponds to PCA performed in a reproducing kernel Hilbert space associated with a positive definite kernel.

Multilinear Generalizations

In multilinear subspace learning, PCA is generalized to multilinear PCA (MPCA) that extracts features directly from tensor representations. MPCA is solved by performing PCA in each mode of the tensor iteratively. MPCA has been applied to face recognition, gait recognition, etc. MPCA is further extended to uncorrelated MPCA, non-negative MPCA and robust MPCA.

Higher Order

N-way principal component analysis may be performed with models such as Tucker decomposition, PARAFAC, multiple factor analysis, co-inertia analysis, STATIS, and DISTATIS.

Robustness – Weighted PCA

While PCA finds the mathematically optimal method (as in minimizing the squared error), it is sensitive to outliers in the data that produce large errors PCA tries to avoid. It therefore is common practice to remove outliers before computing PCA. However, in some contexts, outliers can be difficult to identify. For example in data mining algorithms like correlation clustering, the assignment of points to clusters and outliers is not known beforehand. A recently proposed generalization of PCA based on a weighted PCA increases robustness by assigning different weights to data objects based on their estimated relevancy.

Logistic Regression

In statistics, logistic regression, or logit regression,or logit model, is a type of probabilistic statistical classification model. It is also used to predict a binary response from a binary predictor, used for predicting the outcome of a categorical dependent variable (i.e., a class label) based on one or more predictor variables (features). That is, it is used in estimating the parameters of a qualitative response model. The probabilities describing the possible outcomes of a single trial are modelled, as a function of the explanatory (predictor) variables, using a logistic function. Frequently (and subsequently in this article) "logistic regression" is used to refer specifically to the problem in which the dependent variable is binary—that is, the number of available categories is two—while problems with more than two categories are referred to as multinomial logistic regression or, if the multiple categories are ordered, as ordered logistic regression.

Logistic regression measures the relationship between a categorical dependent variable and one or more independent variables, which are usually (but not necessarily) continuous, by using probability scores as the predicted values of the dependent variable. As such it treats the same set of problems as does probit regression using similar techniques; the former assumed logistic function, the latter standard normal distribution function.

Fields and Example Applications

Logistic regression was put forth in the 1940s as an alternative to Fisher's 1936 classification method, linear discriminant analysis. It is used widely in many fields, including the medical and social sciences. For example, the Trauma and Injury Severity Score (TRISS), which is widely used to predict mortality in injured patients, was originally developed by Boyd et al. using logistic regression. Logistic regression might be used to predict whether a patient has a given disease (e.g. diabetes;coronary heart disease), based on observed characteristics of the patient (age, gender, body mass index, results of various blood tests, etc.;age, blood cholesterol level, systolic blood pressure, relative weight, blood hemoglobin level, smoking (at 3 levels), and abnormal electrocardiogram.). Another example might be to predict whether an American voter will vote Democratic or Republican, based on age, income, gender, race, state of residence, votes in previous elections, etc. The technique can also be used in engineering, especially for predicting the probability of failure of a given process, system or product. It is also used in marketing applications such as prediction of a customer's propensity to purchase a product or cease a subscription, etc. In economics it can be used to predict the likelihood of a person's choosing to be in the labour force, and a business application would be to predict the likehood of a homeowner defaulting on a mortgage. Conditional random fields, an extension of logistic regression to sequential data, are used in natural language processing.

Basics

Logistic regression can be binomial or multinomial. Binomial or binary logistic regression deals with situations in which the observed outcome for a dependent variable can have only two possible types (for example, "dead" vs. "alive"). Multinomial logistic regression deals with situations where the outcome can have three or more possible types (e.g., "disease A" vs. "disease B" vs. "disease C"). In binary logistic regression, the outcome is usually coded as "0" or "1", as this leads to the most straightforward interpretation.

If a particular observed outcome for the dependent variable is the noteworthy possible outcome (referred to as a "success" or a "case") it is usually coded as "1" and the contrary outcome (referred to as a "failure" or a "noncase") as "0". Logistic regression is used to predict the odds of being a case based on the values of the independent variables (predictors). The odds are defined as the probability that a particular outcome is a case divided by the probability that it is a noncase.

Like other forms of regression analysis, logistic regression makes use of one or more predictor variables that may be either continuous or categorical data. Unlike ordinary linear regression, however, logistic regression is used for predicting binary outcomes of the dependent variable (treating the dependent variable as the outcome of a Bernoulli trial) rather than continuous outcomes.

Given this difference, it is necessary that logistic regression take the natural logarithm of the odds of the dependent variable being a case (referred to as the logit or log-odds) to create a continuous criterion as a transformed version of the dependent variable. Thus the logit transformation is referred to as the link function in logistic regression—although the dependent variable in logistic regression is binomial, the logit is the continuous criterion upon which linear regression is conducted.

The logit of success is then fit to the predictors using linear regression analysis. The predicted value of the logit is converted back into predicted odds via the inverse of the natural logarithm, namely the exponential function. Therefore, although the observed dependent variable in logistic regression is a zero-or-one variable, the logistic regression estimates the odds, as a continuous variable, that the dependent variable is a success (a case). In some applications the odds are all that is needed. In others, a specific yes-or-no prediction is needed for whether the dependent variable is or is not a case; this categorical prediction can be based on the computed odds of a success, with predicted odds above some chosen cut-off value being translated into a prediction of a success.

Logistic Function, Odds, Odds Ratio, and Logit

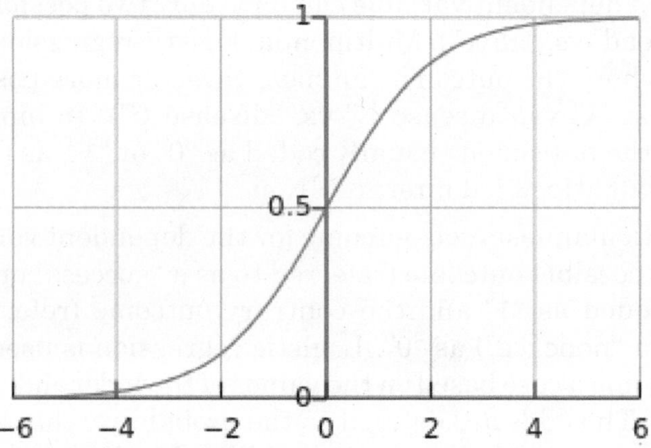

Figure: *The logistic function, with* $\beta_0 + \beta_1 x$ *on the horizontal axis and* $F(x)$ *on the vertical axis*

An explanation of logistic regression begins with an explanation of the logistic function, which always takes on values between zero and one:

$$F(t) = \frac{e^t}{e^t + 1} = \frac{1}{1 + e^{-t}},$$

If t is viewed as a linear function of an explanatory variable x (or of a linear combination of explanatory variables), the logistic function can be written as:

$$F(x) = \frac{1}{1 + e^{-(\beta_0 + \beta_1 x)}}.$$

This will be interpreted as the probability of the dependent variable equalling a "success" or "case" rather than a failure or non-case. It's clear to see that the response variables Y_i are not identically distributed: $P(Y_i = 1 \mid X)$ differs from one data point X_i to another, though they are independent given design matrix X and shared with parameters β. We also define the inverse of the logistic function, the logit (log odds):

$$g(x) = \ln \frac{F(x)}{1 - F(x)} = \beta_0 + \beta_1 x,$$

and equivalently:

$$\frac{F(x)}{1 - F(x)} = e^{\beta_0 + \beta_1 x}.$$

A graph of the logistic function $F(x)$ is shown in Figure above. The input is the value of $\beta_0 + \beta_1 x$ and the output is $F(x)$. The logistic function is useful because it can take an input with any value from negative infinity to positive infinity, whereas the output $F(x)$ is confined to values between 0 and 1 and hence is interpretable as a probability. In the above equations $g(x)$, refers to the logit function of some given linear combination of the predictors, denotes the natural logarithm $F(x)$, is the probability that the dependent variable equals a case, β_0 is the intercept from the linear regression equation (the value of the criterion when the predictor is equal to zero), $\beta_1 x$ is the regression coefficient multiplied by some value of the predictor, and base denotes the exponential function.

The formula for $F(x)$ illustrates that the probability of the dependent variable equaling a case is equal to the value of the logistic function of the linear regression expression. This is important in that it shows that the value of the linear regression expression can vary from negative to positive infinity and yet, after transformation, the resulting expression for the probability $F(x)$ ranges between 0 and 1.

The equation for $F(x)$ illustrates that the logit (i.e., log-odds or natural logarithm of the odds) is equivalent to the linear regression expression. Likewise, the next equation illustrates that the odds of the dependent variable equaling a case is equivalent to the exponential function of the linear regression expression. This illustrates how the logit serves as a link function between the probability and the linear regression expression. Given that the logit ranges between negative infinity and positive infinity, it provides an adequate criterion upon which to conduct linear regression and the logit is easily converted back into the odds.

Odds ratio can be defined as:

$$OR = odds(F(x+1)) / odds(F(x)) = \frac{\dfrac{F(x+1)}{1-F(x+1)}}{\dfrac{F(x)}{1-F(x)}} = e^{\beta_0+\beta_1(x+1)} / e^{\beta_0+\beta_1 x} = e^{\beta_1}$$

or for binary variable $F(0)$ instead of $F(x)$ and $F(1)$ for $F(x+1)$. This exponential relationship provides an interpretation for β1: The odds multiply by e^{β_1} for every 1-unit increase in x.

Multiple Explanatory Variables

If there are multiple explanatory variables, then the above expression $\beta_0 + \beta_1 x$ can be revised to $\beta_0 + \beta_1 x_1 + \beta_2 x_2 + \cdots + \beta_m x_m$. Then when this is used in the equation relating the logged odds of a success to the values of the predictors, the linear regression will be a multiple regression with m explanators; the parameters β_j for all j = 0, 1, 2, ..., m are all estimated.

Model Fitting

Estimation

Maximum Likelihood Estimation: The regression coefficients are usually estimated using maximum likelihood estimation. Unlike linear regression with normally distributed residuals, it is not possible to find a closed-form expression for the coefficient values that maximizes the likelihood function, so an iterative process must be used instead, for example Newton's method. This process begins with a tentative solution, revises it slightly to see if it can be improved, and repeats this revision until improvement is minute, at which point the process is said to have converged.

In some instances the model may not reach convergence. When a model does not converge this indicates that the coefficients are not meaningful because the iterative process was unable to find appropriate solutions. A failure to converge may occur for a number of reasons: having a large proportion of predictors to cases, multicollinearity, sparseness, or complete separation.

- Having a large proportion of variables to cases results in an overly conservative Wald statistic (discussed below) and can lead to nonconvergence.

- Multicollinearity refers to unacceptably high correlations between predictors. As multicollinearity increases, coefficients remain unbiased but standard errors increase and the likelihood of model convergence decreases.

  To detect multicollinearity amongst the predictors, one can conduct a linear regression analysis with the predictors of interest for the sole purpose of examining the tolerance statistic used to assess whether multicollinearity is unacceptably high.

- Sparseness in the data refers to having a large proportion of empty cells (cells with zero counts). Zero cell counts are particularly problematic with categorical predictors. With continuous predictors, the model can infer values for the zero cell counts, but this is not the case with categorical predictors. The model will not converge with zero cell counts for categorical predictors because the natural logarithm of zero is an undefined value, so final solutions to the model cannot be reached. To remedy this problem, researchers may collapse categories in a theoretically meaningful way or add a constant to all cells.

- Another numerical problem that may lead to a lack of convergence is complete separation, which refers to the instance in which the predictors perfectly predict the criterion – all cases are accurately classified. In such instances, one should reexamine the data, as there is likely some kind of error.

As a general rule of thumb, logistic regression models require a minimum of about 10 events per explaining variable (where event denotes the cases belonging to the less frequent category in the dependent variable).

Minimum Chi-Squared Estimator for Grouped Data

While individual data will have a dependent variable with a value of zero or one for every observation, with grouped data one observation is on a group of people who all share the same characteristics (e.g., demographic characteristics); in this case the researcher observes the proportion of people in the group for whom the response variable falls into one category or the other.

If this proportion is neither zero nor one for any group, the minimum chi-squared estimator involves using weighted least squares to estimate a linear model in which the dependent variable is the logit of the proportion: that is, the log of the ratio of the fraction in one group to the fraction in the other group.

Evaluating Goodness of Fit

Goodness of fit in linear regression models is generally measured using the *R*2. Since this has no direct analog in logistic regression, various methods including the following can be used instead.

Deviance and Likelihood Ratio Tests

In linear regression analysis, one is concerned with partitioning variance via the sum of squares calculations – variance in the criterion is essentially divided into variance accounted for by the predictors and residual variance. In logistic regression analysis, deviance is used in lieu of sum of squares calculations. Deviance is analogous to the sum of squares calculations in linear regression and is a measure of the lack of fit to the data in a logistic regression model. Deviance is calculated by comparing a given model with the saturated model – a model with a theoretically perfect fit. This computation is called the likelihood-ratio test:

$$D = -2\ln\frac{\text{likelihood of the fitted model}}{\text{likelihood of the saturated model}}.$$

In the above equation D represents the deviance and ln represents the natural logarithm. The log of the likelihood ratio (the ratio of the fitted model to the saturated model) will produce a negative value, so the product is multiplied by negative two times its natural logarithm to produce a value with an approximate chi-squared distribution. Smaller values indicate better fit as the fitted model deviates less from the saturated model. When assessed upon a chi-square distribution, nonsignificant chi-square values indicate very little unexplained variance and thus, good model fit. Conversely, a significant chi-square value indicates that a significant amount of the variance is unexplained.

Two measures of deviance are particularly important in logistic regression: null deviance and model deviance. The null deviance represents the difference between a model with only the intercept (which means "no predictors") and the saturated model. And, the model deviance represents the difference between a model with at least one predictor and the saturated model.

In this respect, the null model provides a baseline upon which to compare predictor models. Given that deviance is a measure of the difference between a given model and the saturated model, smaller values indicate better fit. Therefore, to assess the contribution of a predictor or set of predictors, one can subtract the model deviance from the null deviance and assess the difference on a χ^2_{s-p}, chi-square distribution with degree of freedom equal to the difference in the number of parameters estimated.

Let

$$D_{\text{null}} = -2\ln\frac{\text{likelihood of null model}}{\text{likelihood of the saturated model}}$$

$$D_{\text{fitted}} = -2\ln\frac{\text{likelihood of fitted model}}{\text{likelihood of the saturated model}}.$$

Then

$$\begin{aligned} D_{\text{fitted}} - D_{\text{null}} &= \left(-2\ln\frac{\text{likelihood of fitted model}}{\text{likelihood of the saturated model}}\right) - \left(-2\ln\frac{\text{likelihood of null model}}{\text{likelihood of the saturated model}}\right) \\ &= -2\left(\ln\frac{\text{likelihood of fitted model}}{\text{likelihood of the saturated model}} - \ln\frac{\text{likelihood of null model}}{\text{likelihood of the saturated model}}\right) \\ &= -2\ln\frac{\left(\frac{\text{likelihood of fitted model}}{\text{likelihood of the saturated model}}\right)}{\left(\frac{\text{likelihood of null model}}{\text{likelihood of the saturated model}}\right)} \\ &= -2\ln\frac{\text{likelihood of the fitted model}}{\text{likelihood of null model}}. \end{aligned}$$

If the model deviance is significantly smaller than the null deviance then one can conclude that the predictor or set of predictors significantly improved model fit.

This is analogous to the F-test used in linear regression analysis to assess the significance of prediction.

Pseudo-R²s

In linear regression the squared multiple correlation, R^2 is used to assess goodness of fit as it represents the proportion of variance in the criterion that is explained by the predictors. In logistic regression analysis, there is no agreed upon analogous measure, but there are several competing measures each with limitations. Three of the most commonly used indices are examined on this page beginning with the likelihood ratio R^2, R^2_L:

$$R_L^2 = \frac{D_{\text{null}} - D_{\text{model}}}{D_{\text{null}}}.$$

This is the most analogous index to the squared multiple correlation in linear regression. It represents the proportional reduction in the deviance wherein the deviance is treated as a measure of variation analogous but not identical to the variance in linear regression analysis. One limitation of the likelihood ratio R^2 is that it is not monotonically related to the odds ratio, meaning that it does not necessarily increase as the odds ratio increases and does not necessarily decrease as the odds ratio decreases.

The Cox and Snell R^2 is an alternative index of goodness of fit related to the R^2 value from linear regression. The Cox and Snell index is problematic as its maximum value is .75, when the variance is at its maximum (.25). The Nagelkerke R^2 provides a correction to the Cox and Snell R^2 so that the maximum value is equal to one. Nevertheless, the Cox and Snell and likelihood ratio R^2s show greater agreement with each other than either does with the Nagelkerke R^2. Of course, this might not be the case for values exceeding .75 as the Cox and Snell index is capped at this value. The likelihood ratio R^2 is often preferred to the alternatives as it is most analogous to R^2 in linear regression, is independent of the base rate (both Cox and Snell and Nagelkerke R^2s increase as the proportion of cases increase from 0 to .5) and varies between 0 and 1.

A word of caution is in order when interpreting pseudo-R^2 statistics. The reason these indices of fit are referred to as pseudo R^2 is because they do not represent the proportionate reduction in error as the R^2 in linear regression does. Linear regression assumes homoscedasticity, that the error variance is the same for all values of the criterion. Logistic regression will always be heteroscedastic – the error variances differ for each value of the predicted score. For each value of the predicted score there would be a different value of the proportionate reduction in error. Therefore, it is inappropriate to think of R^2 as a proportionate reduction in error in a universal sense in logistic regression.

Hosmer–Lemeshow Test

The Hosmer–Lemeshow test uses a test statistic that asymptotically follows a χ^2 distribution to assess whether or not the observed event rates match expected event rates in subgroups of the model population.

Evaluating Binary Classification Performance

If the estimated probabilities are to be used to classify each observation of independent variable values as predicting the category that the dependent variable is found in, the various methods below for judging the model's suitability in out-of-sample forecasting can also be used on the data that were used for estimation—accuracy, precision (also called positive predictive value), recall (also called sensitivity), specificity and negative predictive value. In each of these evaluative methods, an aspect of the model's effectiveness in assigning instances to the correct categories is measured.

Coefficients

After fitting the model, it is likely that researchers will want to examine the contribution of individual predictors. To do so, they will want to examine the regression coefficients. In linear regression, the regression coefficients represent the change in the criterion for each unit change in the predictor.

In logistic regression, however, the regression coefficients represent the change in the logit for each unit change in the predictor. Given that the logit is not intuitive, researchers are likely to focus on a predictor's effect on the exponential function of the regression coefficient – the odds ratio. In linear regression, the significance of a regression coefficient is assessed by computing a t-test. In logistic regression, there are several different tests designed to assess the significance of an individual predictor, most notably the likelihood ratio test and the Wald statistic.

Likelihood Ratio Test

The likelihood-ratio test discussed above to assess model fit is also the recommended procedure to assess the contribution of individual "predictors" to a given model. In the case of a single predictor model, one simply compares the deviance of the predictor model with that of the null model on a chi-square distribution with a single degree of freedom. If the predictor model has a significantly smaller deviance (c.f chi-square using the difference in degrees of freedom of the two models), then one can conclude that there is a significant association between the "predictor" and the outcome. Although some common statistical packages (e.g. SPSS) do provide likelihood ratio test statistics, without this computationally intensive test it would be more difficult to assess the contribution of individual predictors in the multiple logistic regression case. To assess the contribution of individual predictors one can enter the predictors hierarchically, comparing each new model with the previous to determine the contribution of each predictor. (There is considerable debate among statisticians regarding the appropriateness of so-called "stepwise" procedures. They do not preserve the nominal statistical properties and can be very misleading.

Wald Statistic

Alternatively, when assessing the contribution of individual predictors in a given model, one may examine the significance of the Wald statistic. The Wald statistic, analogous to the t-test in linear regression, is used to assess the significance of coefficients. The Wald statistic is the ratio of the square of the regression coefficient to the square of the standard error of the coefficient and is asymptotically distributed as a chi-square distribution.

$$W_j = \frac{B_j^2}{SE_{B_j}^2}$$

Although several statistical packages (e.g., SPSS, SAS) report the Wald statistic to assess the contribution of individual predictors, the Wald statistic has limitations. When the regression coefficient is large, the standard error of the regression coefficient also tends to be large increasing the probability of Type-II error. The Wald statistic also tends to be biased when data are sparse.

Case-Control Sampling

Suppose cases are rare. Then we might wish to sample them more frequently than their prevalence in the population. For example, suppose there is a disease that affects 1 person in 10,000 and to collect our data we need to do a complete physical. It may be too expensive to do thousands of physicals of healthy people in order to get data on only a few diseased individuals. Thus, we may evaluate more diseased individuals. This is also called unbalanced data. As a rule of thumb, sampling controls at a rate of five times the number of cases is sufficient to get enough control data.

If we form a logistic model from such data, if the model is correct, the β_j parameters are all correct except for β_0. We can correct β_0 if we know the true prevalence as follows:

$$\widehat{\beta_0^*} = \widehat{\beta_0} + \log\frac{\pi}{1-\pi} - \log\frac{\tilde{\pi}}{1-\tilde{\pi}}$$

where π is the true prevalence and $\tilde{\pi}$ is the prevalence in the sample.

Formal Mathematical Specification

There are various equivalent specifications of logistic regression, which fit into different types of more general models. These different specifications allow for different sorts of useful generalizations.

Setup

The basic setup of logistic regression is the same as for standard linear regression.

It is assumed that we have a series of N observed data points. Each data point i consists of a set of m explanatory variables $x_{1,i} \ldots x_{m,i}$ (also called independent variables, predictor variables, input variables, features, or attributes), and an associated binary-valued outcome variable Y_i (also known as a dependent variable, response variable, output variable, outcome variable or class variable), i.e. it can assume only the two possible values 0 (often meaning "no" or "failure") or 1 (often meaning "yes" or "success").

The goal of logistic regression is to explain the relationship between the explanatory variables and the outcome, so that an outcome can be predicted for a new set of explanatory variables.

Some examples:

- The observed outcomes are the presence or absence of a given disease (e.g. diabetes) in a set of patients, and the explanatory variables might be characteristics of the patients thought to be pertinent (sex, race, age, blood pressure, body-mass index, etc.).
- The observed outcomes are the votes (e.g. Democratic or Republican) of a set of people in an election, and the explanatory variables are the demographic characteristics of each person (e.g. sex, race, age, income, etc.). In such a case, one of the two outcomes is arbitrarily coded as 1, and the other as 0.

As in linear regression, the outcome variables Y_i are assumed to depend on the explanatory variables $x_{1,i} \dots x_{m,i}$.

Explanatory Variables

As shown above in the above examples, the explanatory variables may be of any type: real-valued, binary, categorical, etc. The main distinction is between continuous variables (such as income, age and blood pressure) and discrete variables (such as sex or race). Discrete variables referring to more than two possible choices are typically coded using dummy variables (or indicator variables), that is, separate explanatory variables taking the value 0 or 1 are created for each possible value of the discrete variable, with a 1 meaning "variable does have the given value" and a 0 meaning "variable does not have that value". For example, a four-way discrete variable of blood type with the possible values "A, B, AB, O" can be converted to four separate two-way dummy variables, "is-A, is-B, is-AB, is-O", where only one of them has the value 1 and all the rest have the value 0.

This allows for separate regression coefficients to be matched for each possible value of the discrete variable. (In a case like this, only three of the four dummy variables are independent of each other, in the sense that once the values of three of the variables are known, the fourth is automatically determined.

Thus, it is only necessary to encode three of the four possibilities as dummy variables. This also means that when all four possibilities are encoded, the overall model is not identifiable in the absence of additional constraints such as a regularization constraint. Theoretically, this could cause problems, but in reality almost all logistic regression models are fit with regularization constraints.)

Outcome Variables

Formally, the outcomes Y_i are described as being Bernoulli-distributed data, where each outcome is determined by an unobserved probability p_i that is specific to the outcome at hand, but related to the explanatory variables. This can be expressed in any of the following equivalent forms:

The meanings of these four lines are:

1. The first line expresses the probability distribution of each Y_i: Conditioned on the explanatory variables, it follows a Bernoulli distribution with parameters p_i, the probability of the outcome of 1 for trial *i*. As noted above, each separate trial has its own probability of success, just as each trial has its own explanatory variables. The probability of success p_i is not observed, only the outcome of an individual Bernoulli trial using that probability.
2. The second line expresses the fact that the expected value of each Y_i is equal to the probability of success p_i, which is a general property of the Bernoulli distribution. In other words, if we run a large number of Bernoulli trials using the same probability of success p_i, then take the average of all the 1 and 0 outcomes, then the result would be close to p_i. This is because doing an average this way simply computes the proportion of successes seen, which we expect to converge to the underlying probability of success.
3. The third line writes out the probability mass function of the Bernoulli distribution, specifying the probability of seeing each of the two possible outcomes.
4. The fourth line is another way of writing the probability mass function, which avoids having to write separate cases and is more convenient for certain types of calculations. This relies on the fact that Y_i can take only the value 0 or 1. In each case, one of the exponents will be 1, "choosing" the value under it, while the other is 0, "canceling out" the value under it. Hence, the outcome is either p_i or $1 - p_i$, as in the previous line.

Linear Predictor Function

The basic idea of logistic regression is to use the mechanism already developed for linear regression by modelling the probability p_i using a linear predictor function, i.e. a linear combination of the explanatory variables and a set of regression coefficients that are specific to the model at hand but the same for all trials. The linear predictor function $f(i)$ for a particular data point i is written as:

$$f(i) = \beta_0 + \beta_1 x_{1,i} + \cdots + \beta_m x_{m,i},$$

where $\beta_0, \ldots, \beta_m$ are regression coefficients indicating the relative effect of a particular explanatory variable on the outcome.

The model is usually put into a more compact form as follows:

- The regression coefficients $\beta_0, \beta_1, \ldots, \beta_m$ are grouped into a single vector β of size $m + 1$.
- For each data point i, an additional explanatory pseudo-variable $x_{0,i}$ is added, with a fixed value of 1, corresponding to the intercept coefficient β_0.
- The resulting explanatory variables $x_{0,i}, x_{1,i}, \ldots, x_{m,i}$ are then grouped into a single vector X_i of size $m + 1$.

This makes it possible to write the linear predictor function as follows:

$$f(i) = \boldsymbol{\beta} \cdot \mathbf{X}_i,$$

using the notation for a dot product between two vectors.

As a Generalized Linear Model

The particular model used by logistic regression, which distinguishes it from standard linear regression and from other types of regression analysis used for binary-valued outcomes, is the way the probability of a particular outcome is linked to the linear predictor function:

$$\operatorname{logit}(\mathbb{E}[Y_i \mid x_{1,i}, \ldots, x_{m,i}]) = \operatorname{logit}(p_i) = \ln\left(\frac{p_i}{1 - p_i}\right) = \beta_0 + \beta_1 x_{1,i} + \cdots + \beta_m x_{m,i}$$

Written using the more compact notation described above, this is:

$$\operatorname{logit}(\mathbb{E}[Y_i \mid \mathbf{X}_i]) = \operatorname{logit}(p_i) = \ln\left(\frac{p_i}{1 - p_i}\right) = \boldsymbol{\beta} \cdot \mathbf{X}_i$$

This formulation expresses logistic regression as a type of generalized linear model, which predicts variables with various types of probability distributions by fitting a linear predictor function of the above form to some sort of arbitrary transformation of the expected value of the variable.

The intuition for transforming using the logit function (the natural log of the odds) was explained above. It also has the practical effect of converting the probability (which is bounded to be between 0 and 1) to a variable that ranges over $(-\infty, +\infty)$ — thereby matching the potential range of the linear prediction function on the right side of the equation.

Note that both the probabilities p_i and the regression coefficients are unobserved, and the means of determining them is not part of the model itself. They are typically determined by some sort of optimization procedure, e.g. maximum likelihood estimation, that finds values that best fit the observed data (i.e. that give the most accurate predictions for the data already observed), usually subject to regularization conditions that seek to exclude unlikely values, e.g. extremely large values for any of the regression coefficients. The use of a regularization condition is equivalent to doing maximum a posteriori (MAP) estimation, an extension of maximum likelihood. (Regularization is most commonly done using a squared regularizing function, which is equivalent to placing a zero-mean Gaussian prior distribution on the coefficients, but other regularizers are also possible.) Whether or not regularization is used, it is usually not possible to find a closed-form solution; instead, an iterative numerical method must be used, such as iteratively reweighted least squares (IRLS) or, more commonly these days, a quasi-Newton method such as the L-BFGS method.

The interpretation of the β_j parameter estimates is as the additive effect on the log of the odds for a unit change in the jth explanatory variable. In the case of a dichotomous explanatory variable, for instance gender, e^{β} is the estimate of the odds of having the outcome for, say, males compared with females.

An equivalent formula uses the inverse of the logit function, which is the logistic function, i.e.:

$$\mathbb{E}[Y_i \mid \mathbf{X}_i] = p_i = \text{logit}^{-1}(\boldsymbol{\beta} \cdot \mathbf{X}_i) = \frac{1}{1 + e^{-\boldsymbol{\beta} \cdot \mathbf{X}_i}}$$

The formula can also be written (somewhat awkwardly) as a probability distribution (specifically, using a probability mass function):

$$\Pr(Y_i = y_i \mid \mathbf{X}_i) = p_i^{y_i}(1 - p_i)^{1-y_i} = \left(\frac{1}{1 + e^{-\boldsymbol{\beta} \cdot \mathbf{X}_i}}\right)^{y_i} \left(1 - \frac{1}{1 + e^{-\boldsymbol{\beta} \cdot \mathbf{X}_i}}\right)^{1-y_i}$$

As a Latent-Variable Model

The above model has an equivalent formulation as a latent-variable model. This formulation is common in the theory of discrete choice models, and makes it easier to extend to certain more complicated models with multiple, correlated choices, as well as to compare logistic regression to the closely related probit model.

Imagine that, for each trial i, there is a continuous latent variable Y_i^* (i.e. an unobserved random variable) that is distributed as follows:

$$Y_i^* = \boldsymbol{\beta} \cdot \mathbf{X}_i + \varepsilon$$

where

$$\varepsilon \sim \text{Logistic}(0,1)$$

i.e. the latent variable can be written directly in terms of the linear predictor function and an additive random error variable that is distributed according to a standard logistic distribution.

Then Y_i can be viewed as an indicator for whether this latent variable is positive:

$$Y_i = \begin{cases} 1 & \text{if } Y_i^* > 0 \ \text{ i.e. } -\varepsilon < \boldsymbol{\beta} \cdot \mathbf{X}_i, \\ 0 & \text{otherwise.} \end{cases}$$

The choice of modelling the error variable specifically with a standard logistic distribution, rather than a general logistic distribution with the location and scale set to arbitrary values, seems restrictive, but in fact it is not. It must be kept in mind that we can choose the regression coefficients ourselves, and very often can use them to offset changes in the parameters of the error variable's distribution. For example, a logistic error-variable distribution with a non-zero location parameter μ (which sets the mean) is equivalent to a distribution with a zero location parameter, where μ has been added to the intercept coefficient. Both situations produce the same value for Y_i^* regardless of settings of explanatory variables.

Similarly, an arbitrary scale parameter s is equivalent to setting the scale parameter to 1 and then dividing all regression coefficients by s. In the latter case, the resulting value of Y_i^* will be smaller by a factor of s than in the former case, for all sets of explanatory variables — but critically, it will always remain on the same side of 0, and hence lead to the same Y_i choice.

(Note that this predicts that the irrelevancy of the scale parameter may not carry over into more complex models where more than two choices are available.)

It turns out that this formulation is exactly equivalent to the preceding one, phrased in terms of the generalized linear model and without any latent variables. This can be shown as follows, using the fact that the cumulative distribution function (CDF) of the standard logistic distribution is the logistic function, which is the inverse of the logit function, i.e.

$$\Pr(\varepsilon < x) = \operatorname{logit}^{-1}(x)$$

Then:

$$\begin{aligned}
\Pr(Y_i = 1 \mid \mathbf{X}_i) &= \Pr(Y_i^* > 0 \mid \mathbf{X}_i) \\
&= \Pr(\boldsymbol{\beta}\cdot\mathbf{X}_i + \varepsilon > 0) \\
&= \Pr(\varepsilon > -\boldsymbol{\beta}\cdot\mathbf{X}_i) \\
&= \Pr(\varepsilon < \boldsymbol{\beta}\cdot\mathbf{X}_i) && \text{(because the logistic distribution is symmetric)} \\
&= \operatorname{logit}^{-1}(\boldsymbol{\beta}\cdot\mathbf{X}_i) \\
&= p_i && \text{(see above)}
\end{aligned}$$

This formulation — which is standard in discrete choice models — makes clear the relationship between logistic regression (the "logit model") and the probit model, which uses an error variable distributed according to a standard normal distribution instead of a standard logistic distribution. Both the logistic and normal distributions are symmetric with a basic unimodal, "bell curve" shape. The only difference is that the logistic distribution has somewhat heavier tails, which means that it is less sensitive to outlying data (and hence somewhat more robust to model mis-specifications or erroneous data).

As a Two-Way Latent-Variable Model

Yet another formulation uses two separate latent variables:

$$\begin{aligned}
Y_i^{0*} &= \boldsymbol{\beta}_0\cdot\mathbf{X}_i + \varepsilon_0 \\
Y_i^{1*} &= \boldsymbol{\beta}_1\cdot\mathbf{X}_i + \varepsilon_1
\end{aligned}$$

where

$$\begin{aligned}
\varepsilon_0 &\sim \mathrm{EV}_1(0,1) \\
\varepsilon_1 &\sim \mathrm{EV}_1(0,1)
\end{aligned}$$

where $\mathrm{EV}_1(0,1)$ is a standard type-1 extreme value distribution: i.e.

$$\Pr(\varepsilon_0 = x) = \Pr(\varepsilon_1 = x) = e^{-x}e^{-e^{-x}}$$

Then

$$Y_i = \begin{cases} 1 & \text{if } Y_i^{1*} > Y_i^{0*}, \\ 0 & \text{otherwise.} \end{cases}$$

This model has a separate latent variable and a separate set of regression coefficients for each possible outcome of the dependent variable. The reason for this separation is that it makes it easy to extend logistic regression to multi-outcome categorical variables, as in the multinomial logit model.

In such a model, it is natural to model each possible outcome using a different set of regression coefficients. It is also possible to motivate each of the separate latent variables as the theoretical utility associated with making the associated choice, and thus motivate logistic regression in terms of utility theory. (In terms of utility theory, a rational actor always chooses the choice with the greatest associated utility.) This is the approach taken by economists when formulating discrete choice models, because it both provides a theoretically strong foundation and facilitates intuitions about the model, which in turn makes it easy to consider various sorts of extensions.

The choice of the type-1 extreme value distribution seems fairly arbitrary, but it makes the mathematics work out, and it may be possible to justify its use through rational choice theory. It turns out that this model is equivalent to the previous model, although this seems non-obvious, since there are now two sets of regression coefficients and error variables, and the error variables have a different distribution. In fact, this model reduces directly to the previous one with the following substitutions:

$$\boldsymbol{\beta} = \boldsymbol{\beta}_1 - \boldsymbol{\beta}_0$$

$$\varepsilon = \varepsilon_1 - \varepsilon_0$$

An intuition for this comes from the fact that, since we choose based on the maximum of two values, only their difference matters, not the exact values — and this effectively removes one degree of freedom. Another critical fact is that the difference of two type-1 extreme-value-distributed variables is a logistic distribution, i.e. if $\varepsilon = \varepsilon_1 - \varepsilon_0 \sim \text{Logistic}(0,1)$.

We can demonstrate the equivalent as follows:

Bayesian Logistic Regression

Comparison of logistic function with a scaled inverse probit function (i.e. the CDF of the normal distribution), comparing $\sigma(x)$ vs. $\Phi(\sqrt{\frac{\pi}{8}}x)$, which makes the slopes the same at the origin. This shows the heavier tails of the logistic distribution.

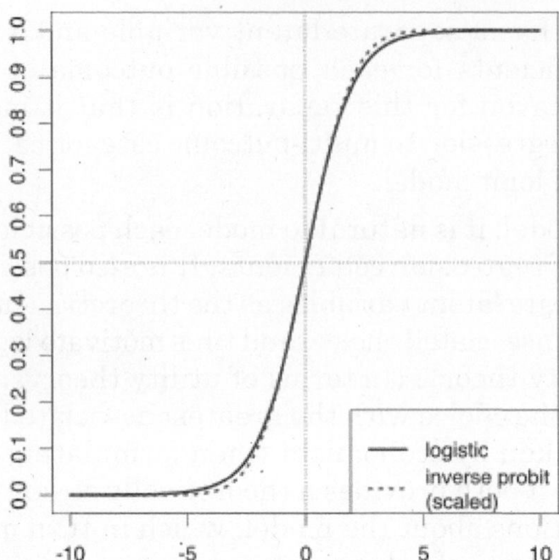

In a Bayesian statistics context, prior distributions are normally placed on the regression coefficients, usually in the form of Gaussian distributions. Unfortunately, the Gaussian distribution is not the conjugate prior of the likelihood function in logistic regression; in fact, the likelihood function is not an exponential family and thus does not have a conjugate prior at all. As a result, the posterior distribution is difficult to calculate, even using standard simulation algorithms (e.g. Gibbs sampling).

There are various possibilities:

- Don't do a proper Bayesian analysis, but simply compute a maximum a posteriori point estimate of the parameters. This is common, for example, in "maximum entropy" classifiers in machine learning.
- Use a more general approximation method such as the Metropolis–Hastings algorithm.
- Draw a Markov chain Monte Carlo sample from the exact posterior by using the Independent Metropolis–Hastings algorithm with heavy-tailed multivariate candidate distribution found by matching the mode and curvature at the mode of the normal approximation to the posterior and then using the Student's t shape with low degrees of freedom. This is shown to have excellent convergence properties.
- Use a latent variable model and approximate the logistic distribution using a more tractable distribution, e.g. a Student's t-distribution or a mixture of normal distributions.

- Do probit regression instead of logistic regression. This is actually a special case of the previous situation, using a normal distribution in place of a Student's t, mixture of normals, etc. This will be less accurate but has the advantage that probit regression is extremely common, and a ready-made Bayesian implementation may already be available.
- Use the Laplace approximation of the posterior distribution. This approximates the posterior with a Gaussian distribution. This is not a terribly good approximation, but it suffices if all that is desired is an estimate of the posterior mean and variance. In such a case, an approximation scheme such as variational Bayes can be used.

Gibbs Sampling with an Approximating Distribution

As shown above, logistic regression is equivalent to a latent variable model with an error variable distributed according to a standard logistic distribution. The overall distribution of the latent variable Y_i* is also a logistic distribution, with the mean equal to $\boldsymbol{\beta}\cdot\mathbf{X}_i$ (i.e. the fixed quantity added to the error variable). This model considerably simplifies the application of techniques such as Gibbs sampling. However, sampling the regression coefficients is still difficult, because of the lack of conjugacy between the normal and logistic distributions. Changing the prior distribution over the regression coefficients is of no help, because the logistic distribution is not in the exponential family and thus has no conjugate prior.

One possibility is to use a more general Markov chain Monte Carlo technique, such as the Metropolis–Hastings algorithm, which can sample arbitrary distributions. Another possibility, however, is to replace the logistic distribution with a similar-shaped distribution that is easier to work with using Gibbs sampling. In fact, the logistic and normal distributions have a similar shape, and thus one possibility is simply to have normally distributed errors. Because the normal distribution is conjugate to itself, sampling the regression coefficients becomes easy. In fact, this model is exactly the model used in probit regression.

However, the normal and logistic distributions differ in that the logistic has heavier tails. As a result, it is more robust to inaccuracies in the underlying model (which are inevitable, in that the model is essentially always an approximation) or to errors in the data. Probit regression loses some of this robustness.

Another alternative is to use errors distributed as a Student's t-distribution. The Student's t-distribution has heavy tails, and is easy to sample from because it is the compound distribution of a normal distribution with variance distributed as an inverse gamma distribution. In other words, if a normal distribution is used for the error variable, and another latent variable, following an inverse gamma distribution, is added corresponding to the variance of this error variable, the marginal distribution of the error variable will follow a Student's t-distribution. Because of the various conjugacy relationships, all variables in this model are easy to sample from.

The Student's t-distribution that best approximates a standard logistic distribution can be determined by matching the moments of the two distributions. The Student's t-distribution has three parameters, and since the skewness of both distributions is always 0, the first four moments can all be matched, using the following equations:

$$\begin{aligned} \mu &= 0 \\ \frac{\nu}{\nu-2}s^2 &= \frac{\pi^2}{3} \\ \frac{6}{\nu-4} &= \frac{6}{5} \end{aligned}$$

This yields the following values:

$$\begin{aligned} \mu &= 0 \\ s &= \sqrt{\frac{7}{9}\frac{\pi^2}{3}} \\ \nu &= 9 \end{aligned}$$

The following graphs compare the standard logistic distribution with the Student's t-distribution that matches the first four moments using the above-determined values, as well as the normal distribution that matches the first two moments. Note how much closer the Student's t-distribution agrees, especially in the tails. Beyond about two standard deviations from the mean, the logistic and normal distributions diverge rapidly, but the logistic and Student's t-distributions don't start diverging significantly until more than 5 standard deviations away.

(Another possibility, also amenable to Gibbs sampling, is to approximate the logistic distribution using a mixture density of normal distributions.)

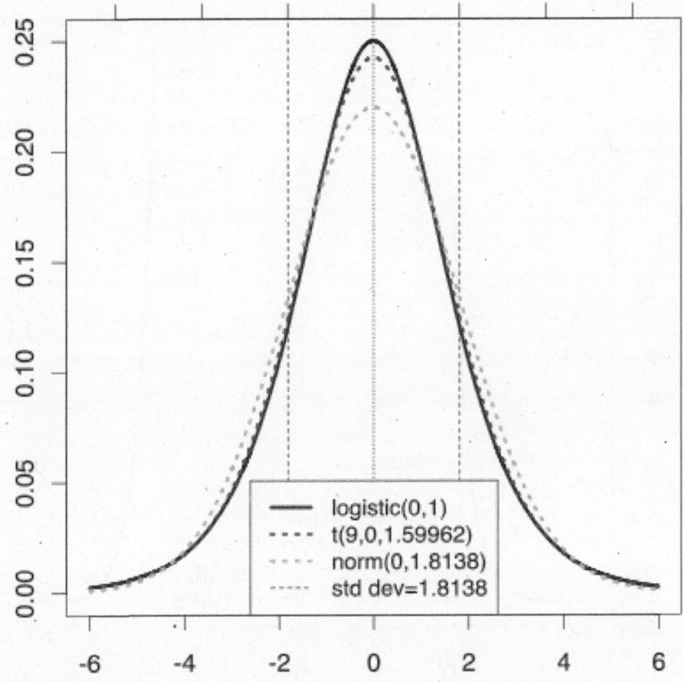

Figure: *Comparison of logistic and approximating distributions (t,normal).*

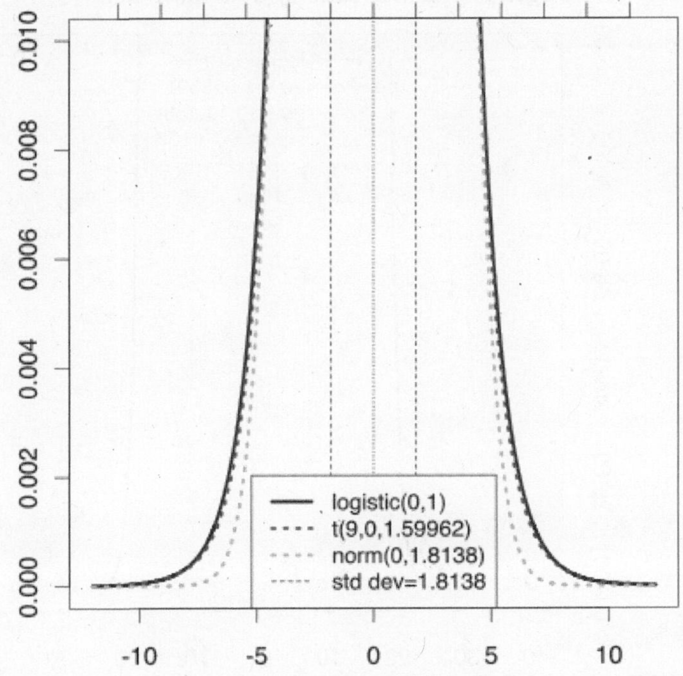

Figure: *Tails of distributions.*

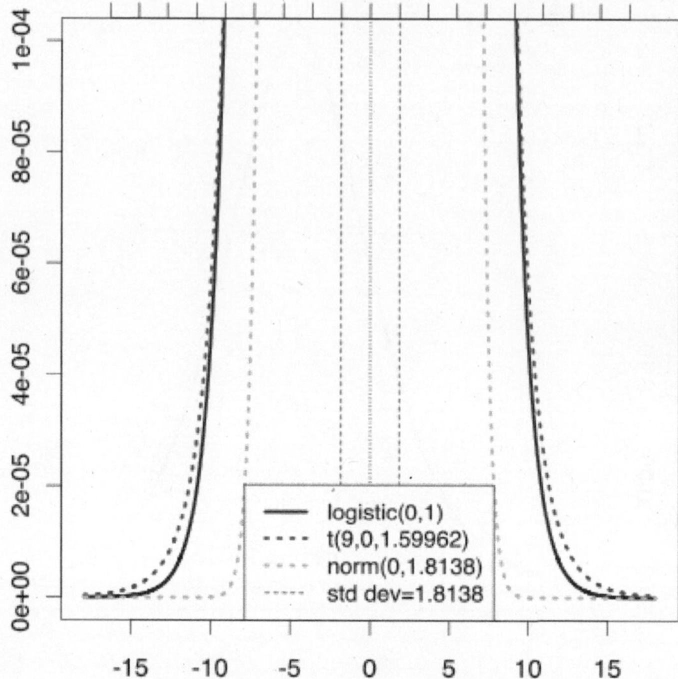

Figure: *Further tails of distributions.*

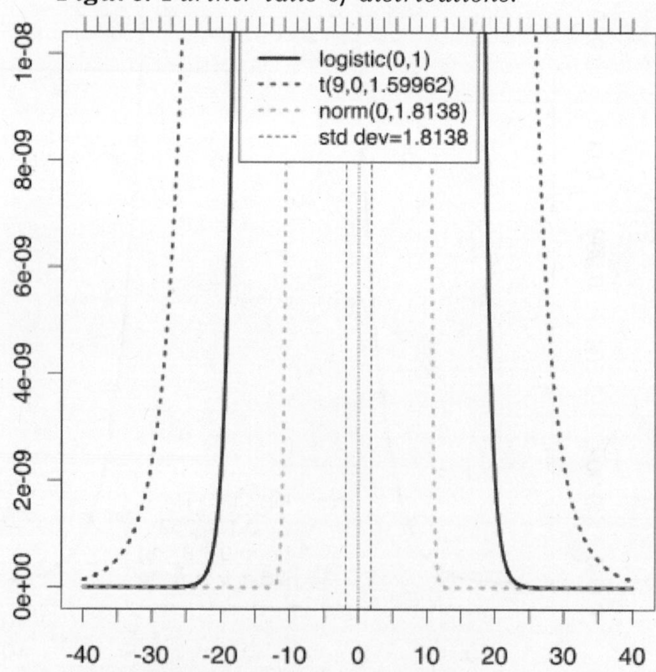

Figure: *Extreme tails of distributions.*

Extensions

There are large numbers of extensions:

- Multinomial logistic regression (or multinomial logit) handles the case of a multi-way categorical dependent variable (with unordered values, also called "classification"). Note that the general case of having dependent variables with more than two values is termed polytomous regression.
- Ordered logistic regression (or ordered logit) handles ordinal dependent variables (ordered values).
- Mixed logit is an extension of multinomial logit that allows for correlations among the choices of the dependent variable.
- An extension of the logistic model to sets of interdependent variables is the conditional random field.

Model Suitability

A way to measure a model's suitability is to assess the model against a set of data that was not used to create the model. The class of techniques is called cross-validation. This holdout model assessment method is particularly valuable when data are collected in different settings (e.g., at different times or places) or when models are assumed to be generalizable.

To measure the suitability of a binary regression model, one can classify both the actual value and the predicted value of each observation as either 0 or 1.

The predicted value of an observation can be set equal to 1 if the estimated probability that the observation equals 1 is above $\frac{1}{2}$, and set equal to 0 if the estimated probability is below $\frac{1}{2}$. Here logistic regression is being used as a binary classification model. There are four possible combined classifications:

1. prediction of 0 when the holdout sample has a 0 (True Negatives, the number of which is TN)
2. prediction of 0 when the holdout sample has a 1 (False Negatives, the number of which is FN)
3. prediction of 1 when the holdout sample has a 0 (False Positives, the number of which is FP)
4. prediction of 1 when the holdout sample has a 1 (True Positives, the number of which is TP)

These classifications are used to calculate accuracy, precision (also called positive predictive value), recall (also called sensitivity), specificity and negative predictive value:

$$\text{Accuracy} = \frac{TP + TN}{TP + FP + FN + TN}$$ = fraction of observations with correct predicted classification

$$\text{Precision} = \text{PositivePredictiveValue} = \frac{TP}{TP + FP}$$ = Fraction of predicted positives that are correct

$$\text{NegativePredictiveValue} = \frac{TN}{TN + FN}$$ = fraction of predicted negatives that are correct

$$\text{Recall} = \text{Sensitivity} = \frac{TP}{TP + FN}$$ = fraction of observations that are actually 1 with a correct predicted classification

$$\text{Specificity} = \frac{TN}{TN + FP}$$ = fraction of observations that are actually 0 with a correct predicted classification

Bibliography

Abraham, Ralph H.; Ueda, Yoshisuke, *The Chaos Avant-Garde: Memoirs of the Early Days of Chaos Theory*, World Scientific, Singapore, 2000.

Alligood, K.T.; Sauer, T.; Yorke, J.A.: *Chaos: an introduction to dynamical systems*, Springer-Verlag, New York, 1997.

Andrienko, N & Andrienko, G.: *Exploratory Analysis of Spatial and Temporal Data*, Springer, New York, 2005.

Badii, R.; Politi A.: *Complexity: hierarchical structures and scaling in physics*: Cambridge University Press, U.S., 1997.

Baker, G. L.: *Chaos, Scattering and Statistical. Mechanics, Cambridge University Press*, U.S., 1996.

Chaos, Arvind Kumar, *Fractals and Self-Organisation*; *New Perspectives on Complexity in Nature*, National Book Trust, New Delhi, 2003.

Collet, Pierre, and Eckmann, Jean-Pierre: *Iterated Maps on the Interval as Dynamical Systems*, Birkhauser, Switzerland, 1980.

Devaney, Robert L.: *An Introduction to Chaotic Dynamical Systems,* Westview Press, United States, 2003.

Elaydi, Saber N.: *Discrete Chaos*, Chapman & Hall/CRC, United Kingdom, 1999.

Gershenfeld, N.: *The Nature of Mathematical Modeling*. Cambridge University Press, New York, 1999.

Gollub, J. P.; Baker, G. L.: *Chaotic dynamics*, Cambridge University Press, London, 1996.

Gutzwiller, Martin: *Chaos in Classical and Quantum Mechanics,* Springer-Verlag, New York, 1990.

Hoaglin, D C; Mosteller, F & Tukey, John Wilder: *Understanding Robust and Exploratory Data Analysis*, Springer, U.S., 1983.

Ivancevic, Vladimir G.; Tijana T.: *Complex nonlinearity: chaos, phase transitions, topology change, and path integrals*, Springer, U.S., 2008.

Kantz, Holger; Thomas, Schreiber: *Nonlinear Time Series Analysis*, Cambridge University Press, London, 2004.

Kautz, Richard: *Chaos: The Science of Predictable Random Motion*, Oxford University Press, London, 2011.

Lorenz, Edward: *The Essence of Chaos*, University of Washington Press, United States, 1996.

M. Mitchell Waldrop, Complexity: *The Emerging Science at the Edge of Order and Chaos*, Simon & Schuster, 1992.

Marshall, Alan: *The Unity of Nature: Wholeness and Disintegration in Ecology and Science*, Imperial College Press, London, 2002.

Martinez, W. L., Martinez, A. R., and Solka, J.: *Exploratory Data Analysis with MATLAB*, Chapman & Hall/CRC, 2010.

Medio, Alfredo; Lines, Marji (2001). Nonlinear Dynamics: A Primer. Cambridge University Press. p. 165. ISBN 0-521-55874-3.

Ott, Edward: *Chaos in Dynamical Systems*, Cambridge University Press London, 2002.

Peitgen, Heinz-Otto and Saupe, Dietmar: *The Science of Fractal Images*, Springer, U.S., 1988.

Peterson, Ivars: *Newton's Clock: Chaos in the Solar System*, Freeman, United States, 1993.

Roulstone, Ian and Norbury, John: *Invisible in the Storm: the role of mathematics in understanding weather*, Princeton University Press, U.S., 2013.

Ruelle, David: *Chaotic Evolution and Strange Attractors*, Cambridge University Press, London, 1989.

Shumway, R. H.: *Applied statistical time series analysis*, Prentice Hall, NJ, 1988.

Sprott, Julien Clinton: *Chaos and Time-Series Analysis*, Oxford University Press, London, 2003.

Theus, M., Urbanek, S.: *Interactive Graphics for Data Analysis: Principles and Examples*, CRC Press, Boca Raton, FL, 2008.

Thompson, J M T, Stewart H B: *Nonlinear Dynamics And Chaos*, John Wiley and Sons Ltd., 2001.

Tufillaro; Reilly: *An experimental approach to nonlinear dynamics and chaos*, Addison-Wesley, 1992.

Velleman, P. F.; Hoaglin, D. C.: *Applications, Basics and Computing of Exploratory Data Analysis*: Addison-Wesley, New York, 1981.

Venables W. N., Ripley, B. D.: *Modern Applied Statistics with S*, Springer, New York, 2002.

Wiggins, Stephen: *Introduction to Applied Dynamical Systems and Chaos*, Springer, New York, 2003.

Zaslavsky, George M.: *Hamiltonian Chaos and Fractional Dynamics*, Oxford University Press, London, 2005.

Index

□□□